Energy

Other volumes in this series:

Energy
Resources, technologies and the environment

Christian Ngô

Preface by Bernard Bigot
French Commissioner for Atomic Energy

Translated from French by
Jeremy Hamand

The Institution of Engineering and Technology

Published by The Institution of Engineering and Technology, London, United Kingdom

Third edition © 2008 Dunod, Paris
English translation © 2010 The Institution of Engineering and Technology

First published 2002
Second edition 2004
Third edition 2008
English translation 2010

The Institution of Engineering and Technology
Michael Faraday House
Six Hills Way, Stevenage
Herts, SG1 2AY, United Kingdom

www.theiet.org

British Library Cataloguing in Publication Data
A catalogue record for this product is available from the British Library

ISBN 978-1-84919-152-4 (paperback)
ISBN 978-1-84919-153-1 (PDF)

Typeset in India by MPS Ltd, A Macmillan Company
Printed in the UK by CPI Antony Rowe, Chippenham

Contents

Preface

Like food and water, energy is a resource essential for life. Its relative abundance has made a crucial contribution to economic and technical development throughout the ages. This development accelerated dramatically during the twentieth century, thanks to the discovery of concentrated and easily exploited sources of energy. The emergence of electricity in the nineteenth century – and its development in the twentieth century – revolutionised the use of energy in many areas: domestic housing and living, distribution and processing of information, transport and so on. However, despite considerable progress, the basic energy needs of large part of humanity remain unmet.

The period of easy and cheap energy for developed countries could end quite soon if we do not make adequate future projections and prepare for a new generation of energy resources. At present over 80% of the world's primary energy production relies on the massive burning of fossil fuels, and we know that, at the pace we are consuming them, we are going to upset the functioning of the planet at a speed and on a scale never experienced under the influence of natural phenomena – unless there is a radical change in our behaviour. And we will soon reach the point at which most fossil fuel resources will be exhausted. We need to act without delay.

The media often focus the attention of the public on energy sources that can only make a marginal impact on energy requirements – and often without sufficient respect for the environment – but give less coverage to those which can make a real difference and offer long-term prospects. It is true that the problem is complex, for no single energy source can meet all our needs. The solutions will depend on many different parameters whose importance will change over time.

Knowing what is at stake, it is a pleasure for me to contribute a preface to this book, which is an excellent introduction to the questions that everyone naturally asks about energy. It provides the necessary foundations for a proper understanding of the subject without overemphasising one energy source over others. The author provides much useful and pertinent data for analysing energy problems, and offers simple, clear and reliable information for the reader to form his own opinion in the light of his own criteria and values.

A clear description of the current energy context and the main sources of energy (fossil fuels, renewable energy and nuclear) is followed by the main uses of energy and an emphasis on energy storage, which is the most difficult problem to resolve. Problems of the impact of energy use on the environment and health are outlined, and the book concludes with a succinct summary of future prospects. This book shows clearly that every energy source has its own advantages and disadvantages, and there is no perfect source of energy that would enable us to meet all our needs, at low cost and without any impact on our environment or health.

We need to avoid all dogmatism and attempt to tackle the problems in the most objective way possible. This book is a step in this direction, and I hope that it proves useful to all those interested in the complex but fascinating subject.

Bernard Bigot
French Commissioner for Atomic Energy

Foreword

To Benjamin, Laurence and Hélène

Energy is present everywhere. It appears whenever there is an interaction between systems, whether living or inert. This is the case with the flea, which expends energy of the order of 10^{-7} J to jump, or human beings, who need around 10^7 J of food per day, heat to warm when they are cold and mechanical energy to multiply their physical strength. It is also the case with natural phenomena that can deploy considerable quantities of energy, such as a Caribbean hurricane $(3.8 \times 10^{18}$ J) [1]. In all phenomena, exchanges of energy are necessary in any dynamic behaviour.

For a large part of history, humans depended on their own physical strength, the strength of animals and the energy that they could derive from firewood, wind and water. The quantities of energy deployed were small and their conditions of living developed only slowly. The massive use of coal enabled them to make a considerable increase in their standard of living. The subsequent exploitation of oil, gas and nuclear energy magnified this phenomenon. Energy thus plays an essential role in economic development. The most spectacular marker of this progress was the more than doubling of life expectancy over 200 years.

This book is an introduction to energy. Its aim is to provide a rapid overview of the whole domain. It is a multidisciplinary subject, and many aspects have to be taken into account: scientific, economic, political, fiscal, environmental, etc. Irrational aspects also intrude, rendering the problem even more complex.

Fossil fuels, renewable energies and nuclear are described in turn after a general introduction to energy. Some uses of energy and its storage are then described, followed by its impact on the environment, for no human activity is entirely anodyne. The final chapter looks at some future perspectives.

Once there are large energy needs – which is the case for a large part of the planet even if all can be satisfied – one quickly realises that there is no good or bad energy. There are only compromises, and many solutions are possible for each situation. Rather than discounting certain solutions, what one should do on the contrary is to try to find a mixture that will best answer the constraints of the problem, which is specific to each country.

Three factors play an important role in energy consumption: technology, which allows consumption to be reduced while providing an equal or better service; governments who through regulation, standard setting and taxation can steer the consumer to particular sources of energy on consumption patterns; and the consumer at the end of the chain, who has a crucial role, for it is they who choose economical or extravagant appliances. What is important is that the consumers have all the information necessary to guide them in their choice and protect our planet, while ensuring their economic development.

I have had a lot of pleasure writing this book, for energy is a fascinating subject, present everywhere throughout our lives. I hope that the readers will also find this pleasure and that this book will help them to form their own opinion on the different sources of energy and its use.

Acknowledgements

Energy is a vast and complex subject, and it is only possible to have an overview after interactions with many experienced people. I have had the opportunity to meet many, from many different disciplines, and I would like to thank them all for the fruitful and enriching discussions which they were kind enough to have with me. These experts came from major research organisations – Commissariat à l'énergie atomique (CEA – French Atomic Energy Commission), Centre national de recherche scientifique (CNRS – French National Centre for Scientific Research), Institut français du pétrole (IFP – French Petroleum Institute), Géosciences pour une terre durable (BRGM – Geoscience for a sustainable earth), industry (EDF, GDF, AREVA, etc.), Agence de l'Environnement et de la Maîtrise de l'Energie (ADEME – French Environment and Energy Management Agency), government ministries, etc. Some are retired and mines of information. I would like to thank the Club Ecrin-énergie which is a meeting place for research laboratories and industrialists where there is always much to be learnt on the subject. Finally, it is above all thanks to the CEA that enabled me to carry out the necessary research to write this book; I would like to thank this excellent organisation whose activities range from fundamental research to the most advanced applied research.

Christian Ngô

Units of energy measurement and conversion factors

Units of energy measurement

1 calorie	4.186 joules
1 kWh	3.6×10^6 joules
1 thermie	4.186×10^6 joules
1 BTU (British Thermal Unit)	1.055×10^3 joules
1 quad	10^{15} BTU
1 barrel of oil	159 litres
1 US gallon	3.785 litres

Conversion factors

1 tonne of oil equivalent (LHV)	42×10^9 joules
1 tonne of coal equivalent (LHV)	29.3×10^9 joules
1000 m^3 natural gas (LHV)	36×10^9 joules
1 tonne liquid natural gas	46×10^9 joules
1000 kWh (primary)	0.0857 TOE (hydro)
1000 kWh (primary)	0.26 TOE (nuclear)
1 tonne natural uranium (PWR)	4.2×10^{14} joules

Chapter 1

Basic concepts

Energy is necessary for all life and all economic activity. Like food and water, energy is indispensable. It has played a fundamental role in the development of civilisations. It has been the cause of war between peoples who have sought throughout history to control access to energy resources. During the twentieth century, easy access to abundant, cheap and concentrated sources of energy enabled faster economic development. The discovery of electricity, a highly convenient carrier of energy, revolutionised the use of energy, and it is almost impossible to imagine a modern house without electricity. However, despite this progress, a considerable proportion of humankind still cannot satisfy its energy needs. Since earliest times, human's energy needs have continually increased. Today we live in a world where we are always needing more, although it is clear that we are reaching the limits of this headlong rush for growth. We need to resolve the thorny problem of how to make further progress without consuming more energy.

Although everyone has a basic idea of what energy is, in reality it is not a notion which is that simple to define. The concept has many faces, and the aspect which interests us here is a particular one that we will now examine more closely.

1.1 What is energy?

Scientists maintain that the fundamental processes (physical phenomena, chemical reactions, biological processes, etc.) that govern our macroscopic world (of objects and living creatures) are ruled by a law in which a physical quantity, which we call *energy*, is always conserved in an isolated system. This is a fundamental law and is associated with the fact that the laws of physics do not change with time. In simple terms, this property translates into the fact that the results of an experiment do not depend on the time it was carried out provided that it was done under exactly the same conditions.[1]

---

[1] Space is also homogenous and isotropic, which leads, respectively, to two laws of conservation: the law of impulsion and the law of kinetic momentum. The homogeneity (isotropy) of space means that an experiment yields the same results if one makes a translation (or a rotation) of the experimental system.

If energy is a physical quantity perfectly defined for the physicist, its dictionary definition is much less clear. But what interests us here pragmatically is that a system or a body possesses energy if it can produce *work* or *heat*.

According to this definition, petrol contains energy since we can use it to propel a vehicle. But this same petrol, when burnt, can provide heat.

At the microscopic level, energy can be seen to exist in either an *organised* or *disorganised* form. In the first case we call it *work*, in the second *heat*. Heat represents the lowest form of energy, because it is distributed over all the degrees of freedom of the system in question, and these are very numerous.[2]

The particularity of energy is to exist in different forms: mechanical, heat, nuclear, etc., and it is very often necessary to convert one form of energy into another. This is done with a certain yield and also loss. When energy passes from a disorganised form (heat) to an organised form (work) the efficiency is rarely good. It is through these transformations from one form to another that humans recover a part of the energy that is exploited for their own needs. It is therefore not the energy contained in a body that is interesting, but that which is acquired through its transformation.

All forms of energy are therefore not of the same quality and cannot be used efficiently to produce work. Thus, a heat source of 300 °C is much more efficient in producing work than a heat source of 50 °C.

The use of energy enables humans to improve their welfare by helping them to feed themselves, keep warm and so on. *Primary energies* are distinguished from *final energies*. *Primary energy* has not undergone any conversion between production and consumption.[3] This is the case with oil, coal, natural gas, hydropower, wood, solar and wind energy. The *final energy* delivered to consumers can be used to meet energy or non-energy needs.

This distinction between primary and secondary energy can have consequences for the evaluation and comparison of different energy sources as can be seen in Figure 1.1. Thus nuclear energy and hydropower produce, on a global scale, much the same quantity of electricity for the consumer. These two sources have the same output, but statistics show us that nuclear energy actually makes three times more primary energy than hydropower (Figure 1.1). The reason is that hydropower is a primary energy and the electricity is produced with an efficiency close to 100%. The electricity produced by nuclear power, which is the energy liberated from uranium fission, on the other hand is not accounted for as a primary energy. Since the efficiency of today's power-generating stations is around 33%, nuclear energy creates three times more primary energy than hydropower, though the energy used by the consumer is the same.

---

[2] In a small glass of water weighing 20 g, the number of degrees of liberty of the water molecules is a multiple of the Avogadro constant ($N = 6.02 \times 10^{23}$). There are therefore more than 10^{24}, that is more than a million billion billion molecules. The heat contained in this water is spread between all these degrees of freedom, and therefore each one possesses very little.

[3] Crude oil is a primary energy, whereas the petrol or diesel obtained by refining are secondary energies. Electricity produced by hydropower or photovoltaic panels is primary, whereas energy of nuclear origin is secondary. Charcoal (secondary energy) is produced from wood (primary energy).

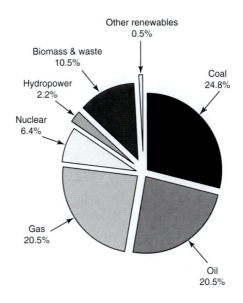

Figure 1.1 *World consumption of primary energy in 2004. Totals do not exactly equal 100% because of rounding errors. [Source: IEA World Energy Outlook 2006]*

1.1.1 Units

The internationally accepted unit of energy is the *joule* (J). For macroscopic transformations this unit is too small so one uses kilojoules (kJ) or megajoules (MJ): 1 kJ = 1000 J and 1 MJ = 10^6 J. Table 1.1 lists the twenty SI prefixes used to form decimal multiples and submultiples of units.

For energy released at the atomic, molecular or nuclear level, the unit used is the *electronvolt* (eV) and its multiples. Its definition is: 1 eV = 1.6×10^{-19} J. Energy released in elementary chemical reactions is of the order of a few eV,

Table 1.1 *SI prefixes*

Prefix	× by	Symbol	Prefix	× by	Symbol
yocto	10^{-24}	y	yotta	10^{24}	Y
zepto	10^{-21}	z	zetta	10^{21}	Z
atto	10^{-18}	a	exa	10^{18}	E
femto	10^{-15}	f	pecta	10^{15}	P
pico	10^{-12}	p	tera	10^{12}	T
nano	10^{-9}	n	giga	10^{9}	G
micro	10^{-6}	μ	mega	10^{6}	M
milli	10^{-3}	m	kilo	10^{3}	k
centi	10^{-2}	c	hecto	10^{2}	h
deci	10^{-1}	d	deca	10^{1}	da

while that released in nuclear reactions is more than a MeV – more than a million times greater. These elementary values may seem small, but there are a large number of elementary reactions in the processes carried out at that scale. For instance, 16 g of methane (1 mol), the basic ingredient of natural gas, contains $N = 6.02 \times 10^{23}$ molecules.

Power is a quantity of energy by unit of time. The basic unit is the *watt* (1 W = 1 J/s). In the energy field, the megawatt (1 MW = 10^6 W), gigawatt (1 GW = 10^9 W) and terawatt (1 TW = 10^{12} W) are often used.[4]

In the field of electricity, the usual energy unit is the watt-hour (Wh) and its multiples. The watt-hour represents an energy of 1 J/s during 1 h. Thus, 1 Wh = 3600 J and 1 kWh = 3.6×10^6 J. The kWh which is a measure of energy must not be confused with the kW which is a unit of power. Also used are the MWh (1 MWh = 10^6 Wh), the GWh (1 GWh = 10^9 Wh) and the TWh (1 TWh = 10^{12} Wh).

1.1.2   Equivalences

To compare different sources of energy it is usual to use as a yardstick the energy provided by crude oil. The conventional unit is the tonne of oil equivalent (TOE), the value of which is fixed at 10^{10} cal (1 cal = 4.18 J) $\simeq$ 42 GJ ($\approx$ 11,700 kWh) [3]. In fact the equivalence 42 GJ $\simeq$ 11,700 kWh corresponds to a direct conversion between the joule and the kWh, with an efficiency of 100%. In practice, for the conversions made the efficiency of the generating station is factored in, which leads to 1 MWh = 0.26 TOE when electricity is produced by a nuclear power station (33% efficiency), 1 MWh = 0.86 TOE for a geothermal power station (10% efficiency) and 1 MWh = 0.086 TOE for electricity produced by a thermal power station or the photovoltaic process, etc. (100% efficiency). This explains why 1000 kWh of electricity represents 0.0857 TOE when produced by hydropower and 0.26 TOE when produced by nuclear reactors, and hence the factor of 3 as mentioned earlier.

When dealing with combustion, one may refer to lower heating value (LHV) or higher heating value (HHV). HHV includes the latent heat of the water vapour produced during combustion whereas the LHV excludes it. As this latent heat is not usually recovered in current processes, LHV is usually used and the TOE is defined according to this convention. The calorific value of crude oil varies slightly between one source and another; it also differs for the various refined petroleum products (1 tonne of petrol = 1.048 TOE, 1 tonne of liquefied petroleum gas (LPG) = 1.095 TOE, 1 tonne of heavy fuel oil = 0.952 TOE [3]).

Coal has a lower calorific value than oil, typically between 0.6 and 0.75 TOE depending on its quality (steam coal, coke, anthracite, etc.). Sometimes the measure of TCE (tonne of coal equivalent) is conventionally set at

---

[4] The *horse power* (1 HP = 736 W) is an old unit introduced by Watt and now out of use. It was based on the work that a vigorous horse could carry out, since the power of one horse power is roughly equivalent to the work of three normal horses [2]. At the end of the nineteenth century, 16,500 horses were needed to work the 38 tramlines of Paris.

1 TCE = 0.697 TOE [3]. Natural gas has a calorific value slightly higher than oil, 1 tonne of liquefied natural gas (LNG) being equivalent to 1.096 TOE (1000 m^3 is equivalent to 0.857 TOE) but 1000 m^3 of natural gas equivalent to 0.857 TOE.

1.2 Energy and development

Energy has always played a major role in human development. Energy consumption increased rapidly following the industrial revolution, particularly during the twentieth century, but this is a much faster process than that which our ancestors experienced in the past. For thousands of years humans were happy with a power of a few hundred watts, at first with their own physical strength then exploiting the strength of their domestic animals. Let us briefly recall the first stages [4].

Human's first energy consumption was naturally food. It enabled them to survive and reproduce. This basic need was extended by other forms of energy that played an increasingly important role in the development of humankind.

Around 500,000 years ago, humans discovered fire and learned to control it. This provided them with light to see at night and scare off wild animals, heat to protect against the cold and to cook their food. About 7000 years ago, humankind invented charcoal that enabled them to make hotter and more efficient fires, and so developed new techniques such as pottery, lead and copper metallurgy, and the manufacture of plaster and lime. Then, about 3000 years ago, they discovered how to smelt and work iron.

As long as humans were hunter-gatherers, their own strength, combined with their intelligence and skills, and fire, were sufficient. But once they started to practise settled agriculture they needed new sources of energy to work the land and mill grain more efficiently. They found this energy in the strength of domestic animals and slaves. The need for slave labour explains the expansion of the Roman Empire, for example. Later, renewable forms of energy largely replaced slaves.

Another form of energy our ancestors needed was for transportation, which is at the heart of our civilisations today. Early means of transport were the backs of human beings and animals. Then sea transport was made possible by using sails to harness wind energy. Coal combustion powered steam engines and petroleum brought the motor car.

While in the developed countries, energy demand, after growing massively, is stabilising and is likely to diminish, thanks to improved *energy efficiency*,[5] in the developing countries it is still growing strongly, as these countries attempt to attain the economic level of the rich countries.

Ten thousand years before the birth of Christ, the world population is believed to have been 5 million, rising to 250 million in 1 AD. The first billion was reached in 1820, and it took 105 years for the population to double to

---

[5] Energy efficiency increases as the same or more work is done with less energy.

2 billion in 1925. Population then increased more rapidly to 3 billion in 1961 and 4 billion in 1976. Six billion was reached in 1999, and in 2009 the population has already reached 6.8 billion. Demographers forecast that, barring catastrophe, world population will reach 8 billion by 2020–25. Projections for the end of the twenty-first century are more uncertain, and population is likely to be smaller than was projected 10 years ago. The most likely scenario is in the range 9–10 billion.[6] Population growth inevitably results in increased demand for energy.

In 2004, world consumption of electricity was 17,400 TWh for a population of 6.4 billion. Average annual consumption is 2700 kWh/person, but this figure does not reflect the real situation in which more than 4 billion people consume less than this value. The domestic electricity consumption of France in 2004 was 477 TWh, of which 32 TWh were lost [5], representing for France's population of some 61 million, an average annual consumption of around 7800 kWh/person.

Life expectancy seems partly correlated to electrical energy consumption because of its impact on countries' standard of living. Life expectancy falls sharply (sometimes as low as 36.5 years) when annual energy consumed is below 1600 kWh/person. At the end of the twentieth century, 3.5 billion people consumed less than 875 kWh/person/year, 2.2 billion of those less than 440 kWh/person, and 1 billion less than 260 kWh/person [6]. The rate of infant mortality also increases sharply when total energy consumption is below ~4400 kWh/person/year [6].

Let us consider some further examples showing the consequences for life expectancy of inequalities in access to energy. Eighty per cent of world energy is consumed by 20% of world population and these people have a life expectancy over 75 years. Sixty per cent consume 19% of energy and their life expectancy is over 50 years. The remaining 20% of world population consume 1% of global energy and their life expectancy is below 40 years.

In 1796, with a population of 28 million, France consumed an average of 0.3 TOE/person/year. Two hundred years later in 1996, consumption had risen to 4.15 TOE/person/year. This was an increase by a factor of 14 per person and 28 for France, because the French population doubled over the period. This corresponds to an average growth in consumption of 1.3% per year, and for France of 1.75% per year. Currently, the global growth in energy consumption is forecast to be 2–2.5% per year. In 200 years, life expectancy in France increased from 27.5 years for men in 1780–89 to 73.5 years in 1994, and for women from 28.1 to 81.8. In 2006, average French life expectancy was 80 years.

The development of the GDP (gross domestic product) per person gives an idea of the prosperity of individuals. In France, between 1400 and 1820, it

---

[6] A world population of 8 billion in 2020 represents an average growth of 1.4% from 2000. If this growth rate was applied over 1000 years, for example, it would result in a total population of 6.5 million billion, which of course is completely unrealistic. An annual growth rate of only 0.2% would result in a total of 50 billion in 1000 years. Conversely, a rate of decline of 0.2% would result in a decline in world population from 6 billion to 810 million in 1000 years.

increased by 0.2% per year that is equivalent to a multiplication of prosperity, over 420 years, of 2.3 times. Since 1950, the increase has been 2.8% per year, a fourfold multiplication of prosperity in 50 years.

Energy should not be wasted, because while it is cheap today,[7] it is likely that this will change in the future. We must prepare for tomorrow by developing every possible source of energy, taking into account the economic, political, security of supply and environmental aspects. In particular, we must apply the true costs of energy that include what economists call the externalities (pollution, greenhouse effect, decommissioning, etc.). These are generally not taken into account, except in rare cases like nuclear energy.

1.3   The Sun

The Sun is a spherical star which is our source of life, for it provides most of the energy that we use today. In fact, apart from geothermal and nuclear, all energy comes from the Sun.

The Sun has a radius of 696,000 km and a mass of the order of 1.99×10^{30} kg. Its surface temperature is 5780 K.[8] According to the Standard Solar Model, accepted by the whole scientific community, the Sun's temperature increases considerably beneath the surface, and reaches 15.6 million degrees in the centre. The Sun consists (by mass) of 71% of hydrogen,[9] 27% of helium and 2% of heavy elements such as carbon, oxygen and iron. The density and pressure at the centre are, respectively, 148,000 kg/m^3 and 2.29×10^{11} times atmospheric pressure.

The Sun was formed 4.55 billion years ago by the gravitational contraction of a cloud of hydrogen, helium and traces of other elements [7]. This process was rapid until the atoms of the cloud were ionised. The energy could no longer escape from this cloud and it slowly contracted. Half of the gravitational energy liberated was converted into radiation and the other half served to heat the cloud. The contraction continued and the cloud heated up. When the temperature was close to 1 million Kelvin, thermonuclear fusion reactions between hydrogen and the light elements such as deuterium, lithium, beryllium and boron began. As the light elements were present in small quantities, the liberation of energy was limited, but was enough to form a gas of very high temperature and set off the fusion reactions between the large numbers of hydrogen atoms, or, to be more precise, the protons.

Most of the thermonuclear reactions took place in the centre of the Sun in a volume corresponding to a sphere whose radius was only about 20% of the Sun's [8]. They provided the Sun with energy and led to the formation of nuclei of helium, ^{4}He, which is a particularly stable element. In the Sun, hydrogen is

---

[7] A 'barrel' of some mineral waters costs around $140, or twice as much as crude oil at $70 a barrel.
[8] This temperature means that a large part of the Sun's radiation is in the visible spectrum. The energy radiated by the Sun is around 4×10^{26} W.
[9] Or more than 90% in terms of the number of atoms.

consumed in fusion reactions, the first stage of which is the interaction between two protons.[10] Altogether four protons are needed to generate one nucleus of helium. These fusion reactions are classified into three families. The first[11] occurs in 85% of cases and liberates 26.2 MeV. The second family (15% of cases) liberates 25.7 MeV and the third (in only 0.2% of cases) liberates 19.1 MeV.

The average energy liberated by a proton in a fusion reaction is 15 MeV. The first reaction of each family corresponds to a reaction between two protons. The probability that the two will react is very small; among the 3×10^{31} m^{-3} protons present, only 5×10^{13} m^{-3}/s lead to fusion. The energy liberated is 120 W/m^3, ten times less than the energy the human body needs to survive ($\simeq 1400$ W/m^3) [7].

During fusion, it is the first reaction between two protons which is the slowest and which governs the process. Some 5 billion years are needed for one proton (^{1}H) to fuse with another proton [7] to form a deuteron (^{2}H). But only about one second is needed for the deuteron formed during this reaction to react with a proton to create a ^{3}He. Around 300,000 years are needed for two ^{3}He to meet and form a nucleus of helium, ^{4}He. This low probability of reaction means that the density of the energy emitted in space by the Sun is very weak: 200 nW/g, or 7000 times less than the energy released by a human being's metabolism ($\simeq 1.4$ mW/g).[12]

For the Sun, hydrogen is not a renewable energy. This fuel will gradually run out, and in about 5 billion years, the Sun will become a 'red giant'. The central part of the core will contract and heat up until the temperature and density of matter reaches a point at which thermonuclear fusion of the helium is triggered. At the same time, the external part of the Sun will expand to form a 'red giant'. The Earth will be destroyed during this expansion [9].

1.4   Energy consumption

The various sources of primary energy that we can use are fossil and mineral resources (coal, oil, gas, uranium) and renewable energies (hydro, solar, wind, biomass, geothermal). The problem with many of these sources is their availability and cost.

Until 200 years ago, humans only used renewable energies: wood for heating and cooking, animal traction for transport, water and wind power for

---

[10] When fusion reactions between neutrinos (ν_e) and photons (γ) are created, the probability of a neutron formed in the centre of the Sun escaping from the Sun without interaction is 10^{-9}, which means that only a single neutrino out of a billion interacts before escaping. It is much more difficult on the other hand for a photon to escape from the Sun on account of the successive interactions which it undergoes. A photon created within the Sun will take some 50,000 years to escape from it.

[11] ^{1}H + ^{1}H → ^{2}H + e$^+$ + ν_e
^{1}H + ^{2}H → ^{3}He + γ
^{3}He + ^{3}He → ^{4}He + 2 ^{1}H

[12] The metabolism of a child, which needs more energy than adult metabolism, is around 3 mW/g. The metabolism of a bacterium can reach 100 W/g [1].

mechanical energy. During the nineteenth century, coal mining led to the development of steam engines. In the twentieth century, oil, gas and nuclear energy were exploited.

Total energy consumption worldwide (commercial and non-commercial) in 2004 was 11.2 GTOE. Fossil fuels (oil, coal, gas) covered more than 80% of requirements (see Table 1.2). The predominance of fossil fuels can also be seen in Figure 1.1.

The growth of primary energy consumption in France between 1960 and 2000 is shown in Figure 1.2. During this period, consumption tripled, representing an annual growth 2.8%. In 2005, consumption reached 276.3 MTOE. However, huge losses take place between the primary and the final energy used by the consumer. Thus, in France, the final energy consumed in 2005 was 160.6 MTOE,

Table 1.2   Consumption of primary commercial energy worldwide in 2004

Energy	GTOE	%
Oil	3.940	35.2
Gas	2.302	20.5
Coal	2.773	24.8
Nuclear	0.714	6.4
Hydro	0.242	2.2
Biomass & waste	1.176	10.5
Other renewables	0.057	0.5
Total	**11.204**	**100**

Totals do not exactly equal 100% because of rounding errors (Source: IEA World Energy Outlook 2006).

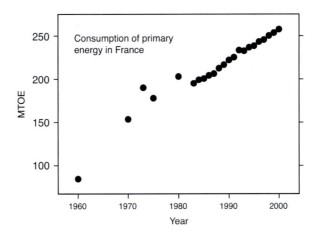

Figure 1.2   Development of primary energy consumption in France [5]

only 68% of primary energy. The distribution of energy consumption in France in 2008 across different energy sources is shown in Table 1.3 [11].

Table 1.3   Consumption of primary energy in France in 2008

Energy source	MTOE	%
Coal	12	4.4
Oil	89	32.5
Gas	41	15.0
Electricity (nuclear and renewables)	117	42.7
Thermal renewable energy	15	5.5
Total	**274**	**100**

Totals do not exactly equal 100% because of rounding errors (Source: DGEMP).

Electricity is used the most widely. Figure 1.3 shows the development of domestic electricity consumption in France up to 2000 [5]. It increased by a factor of 6.125 between 1960 and 2000, an annual growth rate of 4.64%. Between 1970 and 2000, consumption rose from 140 to 441 TWh (482 TWh in 2005). During the 1970s, it was therefore necessary to find new means of generating electricity. In 1960, large hydropower stations produced 56% of French electricity but virtually all sites were already exploited. Coal and oil-fired power stations were developed until the 1973 oil price shock which led to the all-out development of nuclear energy. That improved the French balance of payments because oil would have to be paid for in foreign currency (some €1000 per head of the population would have been needed to buy the necessary oil at $80 a barrel to produce the necessary electricity in oil-fired power stations).

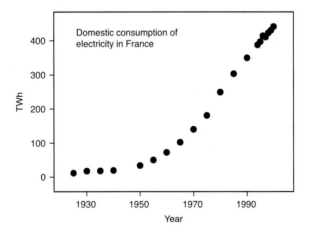

Figure 1.3   Development of domestic consumption of electricity in France [5]

Table 1.4 shows the distribution of total final energy and electricity consumption in 2005. The total final energy consumption is 160.7 MTOE as against 276.2 MTOE of total primary energy consumed in 2005.

Table 1.4   Distribution of the consumption of final energy and electricity by economic sector

Sector	Total final energy		Electricity	
	MTOE	%	TWh	%
Industry	37.7	23.4	135.8	32.0
Agriculture	2.9	1.8	3.4	0.8
Residential, tertiary	69.8	43.4	246.4	64.3
Transport	50.4	31.3	10.4	2.8
Total	**160.7**	**100**	**423.7**	**100**

Totals do not exactly equal 100% because of rounding errors (Source: [5, 11]).

In the next century, the world will be confronted with two fundamental problems. The first relates to cheap fossil fuel reserves and the second to the greenhouse effect.

1.5   The greenhouse effect

Without the greenhouse effect the average temperature of our planet would be -18 °C. With this phenomenon it is 15 °C. This represents an average energy difference of 150 W/m^2.[13] Since the beginning of the pre-industrial era

---

[13] At the entry to the Earth's atmosphere, perpendicular to the Earth–Sun axis, the energy received from the Sun actually averages 1367 W/m^2, a value which is called the *solar constant*. The average intensity received on Earth is calculated bearing in mind that while the surface of the Earth, a sphere of radius R, equals $4\pi R^2$, the Sun only sees a disk with a surface of πR^2. The average energy received on Earth therefore equals a quarter ($\pi R^2/4\pi R^2 = 1/4$) of the solar constant, about 340 W/m^2.

The radiation balance of our planet is in equilibrium with the Sun because the energy received from the Sun is equal to that emitted by Earth. To calculate this, we use Stephan's Law which says that the energy ε emitted by surface unit of a body brought to temperature T is $\varepsilon = \sigma T^4$, or roughly $\varepsilon = \left(\frac{T}{64.5}\right)^4$ W/m^2, where σ is a constant ($\sigma = 5.674 \times 10^{-8}$ Wm^{-2} K^{-4}). Of the average 340 W/m^2 that arrive from the Sun, nearly 30% are reflected (100 W/m^2) back into space and 70% (240 W/m^2) are absorbed by our planet. Of these 240 W/m^2, 70 W/m^2 (around 20%) are absorbed by the atmosphere which is warmed and the rest (170 W/m^2, or 50%) heats the continents and oceans.

If the temperature of the Earth was -18 °C, or 255.16 K, it would emit 240 W/m^2. Thanks to the greenhouse effect, the average temperature is $+15$ °C which results in an emission of 390 W/m^2. As 240 W/m^2 must be emitted into space to achieve the Earth–Sun balance, 150 W/m^2 must be absorbed by the greenhouse effect of the atmosphere.

the greenhouse effect has increased by 2.45 W/m^2, or almost 1% of the energy emitted by our planet. The effect of this has been to raise the average temperature, between 1850 and 1995, by around half a degree. This increase is worrying.

Water vapour is the gas with the biggest greenhouse effect (60–70% of the total). But the amount generated by humans do not have a significant effect on its concentration in the atmosphere, and the water cycle is very rapid [10]. This is not the case with other gases like carbon dioxide (CO_2), methane (CH_4) and nitrous oxide (N_2O). The halogen gases (CFC, halons, etc.[14]) are emitted in smaller quantities and their impact is minor, but their persistence is much greater. These halogen gases also play an important role in the destruction of the ozone layer that protects us from the harmful ultraviolet radiation. Measures were taken at an international level to limit their use (Vienna Convention 1985, Montreal Protocol 1987), but it will still take several decades to restore the ozone layer to its 1970s level.

The gases have different levels of greenhouse effect, which can be quantified as follows: 1.56 Wm^{-2} for CO_2, 0.5 Wm^{-2} for CH_4, 0.14 Wm^{-2} for N_2O and 0.25 Wm^{-2} for the CFCs. All fossil fuels emit CO_2 when burned, as they contain carbon. Better management of fuel combustion and the choice of fossil fuel (e.g. for the same quantity of energy generated, the burning of natural gas emits around half the CO_2 of coal) can optimise greenhouse gas emission, but it is impossible to reduce it to zero because the combustion of carbon compounds always yields carbon dioxide. By contrast, the production of renewable and nuclear energy does not contribute to an increase in the greenhouse effect.

The increase in man-made greenhouse gas emissions could have serious consequences on the environment according to a number of predictive models [10]. A number of scenarios have been developed to estimate the average temperature in 2100. They point to an average warming of between 2 °C and 6 °C. The higher values would have dramatic effects on the environment, including a notable rise in sea level, and the appearance of tropical diseases in some countries that do not have them today [10]. The Intergovernmental Panel on Climate Change (IPCC) also predicts on the basis of models that sea levels could rise by between 15 and 95 cm in 100 years, 95% of European glaciers will disappear, precipitation patterns will change (with heavier rainfall in Europe) and major climatic disturbances (cyclones, hurricanes, tornadoes, etc.) will become more frequent. These predictions are sufficiently worrying for attempts to be made at a global level for an international agreement on limiting greenhouse gas emissions. The Kyoto Conference in December 1997 made a start in this direction, although many experts consider that it did not go far enough.

---

[14] The CFCs (chlorofluorocarbons) are carbon compounds in which hydrogen atoms are replaced by atoms of chlorine or fluorine. In halons, some atoms of hydrogen are replaced by atoms of bromine and/or fluorine.

1.6   Conclusion

Two factors will lead to an increased demand for energy in the future: the growth of world population and the fact that the developing countries aspire to increase their standard of living. If we assume an annual global increase of 2–2.5% in energy demand, the world's energy consumption will double within 30 years. To meet these additional needs, without adding too much to the greenhouse effect, it will be necessary to develop nuclear and renewable energies, which currently only account for 20% of the world's energy consumption.

All energy sources have their own advantages and disadvantages, in terms of cost, security of supply, impact on the environment, etc. There is no universal solution, and the best mix of energy sources will vary from country to country.

The consumption of primary energy will remain for several decades largely dominated by fossil fuel combustion, especially oil. Fossil fuels represent nearly 90% of commercial energy (80% if one includes non-commercial energy) and nothing else can replace them quantitatively or economically. Between the beginning and the end of the twentieth century, the world consumption of primary energy rose from around 1 GTOE to tens of GTOE. This is what enabled humanity to make such major strides in economic development over that period.

Chapter 2
Fossil fuels

The fossil fuels, comprising coal, oil and gas, were formed from vegetable or animal living matter. They contain carbon that releases energy and carbon dioxide when burnt. The quantity of carbon contained in fossil fuels only represents a very small proportion of the carbon existing on the Earth [13]. It is estimated that the vast majority of this carbon (99.75% or 2×10^7 billion tonnes) is stored in the form of carbonates in sediments. The rest, around 0.25% (50,000 billion tonnes), enters the carbon cycle linked to living matter. Part of the 50,000 billion tonnes of carbon exists in the form of carbonic gas in the deep oceans (34,500 billion tonnes) or fossil carbon (10,000 billion tonnes). Around 1200 billion tonnes of carbon are in the form of atmospheric carbon dioxide. Terrestrial and marine life represent around 1150 and 3000 billion tonnes of carbon, respectively.

2.1 Coal

Coal was formed in the Carboniferous Era, between 345 and 280 million years ago, from vegetable matter flooded after major geological upheavals. The vegetation developed using radiated solar energy slowly decomposed in anaerobic conditions and turned into coal, which enabled the storage of this energy. A layer of vegetation 1000 m thick was slowly transformed into a 50-metre seam of coal [14]. This phenomenon, which occurred in several regions of the world, was repeated several times leading to seams of coal separated by layers of sedimentary rock.

The Chinese were already using coal to fire porcelain in 1000 BC [15]. It was not used in Europe until much later, and then with some hesitation. Around the twelfth century, a few people began burning coal because wood was becoming rare and expensive due to its massive use for heating and for house and ship building. Coal was disliked because it was dirty, caused air pollution and affected the lungs. As it smelled of sulphur when burnt due to the sulphuric compounds in its smoke, those who used it were associated with witchcraft. It was even banned in London by the King of England because of the air pollution it caused. The gradual reduction of forest cover meant that by the seventeenth century coal was accepted as a fuel: it represented the only alternative to wood to ensure economic development. It played an important role in the steel industry during the first half of the eighteenth century. At the beginning of the nineteenth century, coal gas was introduced and coal was the main fuel used for firing steam engines.

Coal contains volatile organic material and carbon in variable proportions. The older it is, the richer in carbon. The volatile compounds burn more easily but produce less heat than carbon, which on the other hand is more difficult to burn.

The richest and oldest coal is *anthracite*, which contains 90–95% of carbon. It is mined, for example, in Russia. Although it is difficult to ignite, it is an excellent fuel that gives off plenty of heat (8.4–9.1 kWh/kg).

Ordinary coal has a lower carbon content, ranging from 70% to 90%. Its calorific value is less, but it is cheaper. It ignites more easily since it contains more volatile compounds. Coal was used for many decades to produce town gas, a blend of carbon monoxide and hydrogen.

Lignite or brown coal has a lower carbon content, below 50%. It is a young coal, formed during the Secondary and Tertiary Eras, found in larger quantities in Germany. It is extracted by open-cast mining and on combustion generates considerable pollution.

Peat is the poorest fossil fuel. Formed in the Quaternary Era, it contains little carbon and causes extreme pollution. Much of it is found in Ireland.

For individual and collective heating systems, the best coal is anthracite. *Coal nuts* made from heating a mixture of coal dust and a binder-like bitumen, are also used. Originally they were highly polluting, but now they are made 'smokeless' by heating to 350 °C.

Flame coal, named because it burns with long flames, is a good variety for producing electricity. It is found in Poland, South Africa and Australia.

Open-cast mines are easy to exploit. Deep mines have much higher investment and exploitation costs. The average depth of coalmines varies from one region to another, from 100 m in the Appalachians to 400 m in France and 700 m in the Ruhr [14]. At the beginning of the twentieth century, yields from mines that were easy to exploit were five times those from mines that were hard to exploit. Today, technology has increased the gap, with yields ranging from 33 tonnes per man per day in open-cast mines in the United States and Australia against 1 tonne in some deep mines in China and India [14].

At the beginning of the twentieth century, coal was the dominant fossil fuel, representing more than half the global consumption of primary energy [14]. Oil represented 2%, natural gas 1%, and wood and other energies nearly a quarter. Today, coal is still important but is no longer the first source of energy. In 2005, it only represented 25% of global primary energy consumption.

Figure 2.1 shows coal, oil and gas consumption in France between 1960 and 2000. The graph shows a slow decline in coal consumption, while natural gas consumption increases over the same period.

France extracted some 60.3 Mt (million tonnes) of coal in 1958. Since then, production has fallen consistently and by 2000, only 4.1 Mt was mined. In

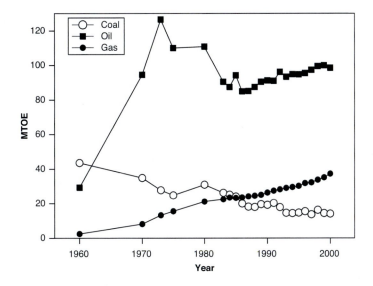

*Figure 2.1   Coal, oil and gas consumption in France between 1960 and 2000
[Source: French Atomic Energy Commission (CEA), 2001]*

1999, Germany, which has substantial coal reserves, still produced 51% of its electricity from coal.

Considerable progress has been made in coal combustion, notably through the fluidised bed technique. This enables the combustion yield to be increased and emissions to be reduced. Coal rich in ash or with high sulphur levels can be used with devices that reduce pollution. A typical coal-fired power station of 1000 MW output, which produces 6.6 TWh of electricity per year (6600 h of production), consumes 6.5 Mt of oxygen, 2.52 Mt of carbon, and emits 7.8 Mt of CO_2, 40,000 tonnes of SO_2, 9500 tonnes of NO_2 and 6000 tonnes of dust. It produces 450,000 tonnes of solid waste [16]. It also emits a substantial amount of radioactivity because coal contains natural radioactive isotopes.

The problem of coal's radioactivity is not considered very often. It is far from negligible and there follow some estimates taken from Reference 17. Coal contains uranium and thorium, the latter in quantities two and a half times higher. The concentration of uranium can range from 1 to 10 parts per million (ppm) with an average of 1.3 ppm for uranium and 3.2 ppm for thorium. It is estimated that in 1982, the United States emitted 100 tonnes of uranium and 1971 tonnes of thorium. At the global level, 3640 tonnes of uranium and 8960 tonnes of thorium were emitted in the course of that year. Uranium and thorium have radium, radon, etc., as decay products, which are also radioactive. For power stations of equal output, the health impact of the radioactivity emitted by a coal-fired power station is 100 times higher than that of a nuclear power station. The average radioactivity of coal is 156,000 becquerels per tonne.[1] The impact of this radioactivity on man is however negligible compared with natural background radioactivity. The nuclear energy potentially contained in the uranium and thorium is superior to the energy liberated by the combustion of coal [17].

Techniques of *gasification* or *liquefaction* can be applied to coal. In the case of liquefaction, the aim is to synthesise liquid fuels but this is not yet economically competitive. As coal reserves are much larger than oil or gas reserves, these methods may prove useful as fossil fuel reserves become increasingly rare.

2.2   Oil

The formation of oil and gas is a complex process with a low energy output. Oil is formed from plankton deposited on the ocean bed – organic matter mixed with sediment accumulated in layers at great depth. The yield is low because 99.9% of the carbon is recycled in the food chain. As water is low in oxygen, the organic matter trapped is transformed by anaerobic microorganisms into carbon macromolecules called *kerogen*. Under the effects of pressure and temperature kerogen breaks down, with nitrogen and oxygen being eliminated to leave hydrocarbon compounds, mainly crude oil. The higher the cracking temperature, the more short carbon chains are obtained.[2] These are found initially in what is called *bedrock*. When the bedrock is fractured under high pressure, microscopic droplets of oil are expelled, rising and migrating to reservoir rock. The yield of this migration is low, since 98% of the oil remains in the bedrock [18]. Gas migrates to the top part by diffusion, while water remains in the lower part and oil between the two. A layer of oil is not a lake but a layer of porous rock containing oil in its microscopic fissures. If the reservoir rock is covered by an impermeable layer, a deposit of oil is formed. The dispersion of droplets of oil throughout the porous rock means that all the oil can never be extracted from an oilfield. Normally some 25–30% is recovered[3] but new techniques enable up to 40% or even more to be achieved. Oil exploration is very expensive. The cost of an exploratory well ranges from 500,000 to 100 million euros (15 million euros on average). Between 20,000 and 100,000 exploratory drillings are made every year for oil [18]. Of the 30,000 oil-bearing strata discovered, 1% contain 75% of the known reserves [18]. The well's flow rate is also a prime consideration. The average flow rate of oil wells in the United States is 0.04 l/s whereas it is 7 l/s in the Middle East [14].

Oil was already being used 3000 years ago in Mesopotamia where bitumen oozing from the ground was collected and used for making mortar, caulking

---

[1] One becquerel is equal to one radioactive decay per second.
[2] The cracking takes place at depths of between 1000 (60 °C) and 4000 m (110–120 °C) and takes tens of millions of years. At greater depths where temperature (120–140 °C) and pressure are higher (300–1500 bar), cracking only yields methane [18].
[3] Using only the natural pressure of the deposit, the recovery rate is only round 10%. With assisted recovery, some 30% is achieved.

boats, lining cisterns, etc. The first oil well, sunk in 1859, marked the beginning of the rush for 'black gold'. The Standard Oil Company, formed by D. Rockefeller in 1870, controlled 80% of production until 1911, when it was dismantled under American anti-trust legislation and split into 34 different companies. The price of oil had remained stable for several decades, before the price rose by a multiple of five times between 1916 and 1920. Then, from the 1930s, around 40 years the price of a barrel of oil remained stable between $1 and $2. Oil was gradually recognised as the most convenient source of energy. As a liquid it is easily transportable, while coal is solid and less compact and gas requires large storage volumes. During the first 'oil shock' in 1973, the price of oil quadrupled. This shows that it is not a good thing that a major part of world oil production should be controlled by a small number of players, a consideration stressed by recent events in 2008–09, when oil prices fluctuated wildly between over $140 and under $40 a barrel. In 2000, the world production of crude oil was 3.5 Gt (gigatonnes) of which 1.4 Gt came from the OPEC countries. In that year, Saudi Arabia was the biggest producer with 402 Mt (11.8%), followed by Russia (316 Mt) and the United States (290 Mt). Western Europe produced 320 Mt (9.6%), 89% of which came from the United Kingdom and Norway. French oil production was very less, 1.4 Mt in 2000, less than 1.5% of national requirements. Figure 2.1 shows that French oil consumption increased sharply until 1970–75 before declining following the commissioning of nuclear power stations. Growth has subsequently been modest. In 2000, oil still made up 40% of French energy requirements. In 2006, it made up 33% but this is because part of national consumption has shifted to natural gas.

Crude oil contains a large number of compounds that are separated by distillation into fractions during the refining process. This produces gases $(C_1–C_4)$,[4] naphtha $(C_5–C_7)$, petrol or gasoline $(C_6–C_{12})$, kerosene $(C_{12}–C_{15})$, diesel $(C_{14}–C_{17})$, etc. Global refining capacity in 2000 was 4 GTOE, which was in excess of requirements but the situation now is more problematic.

Around 1.5 billion tonnes of crude oil and 0.5 billion tonnes of refined products are transported by sea, making up 40% of the world's maritime freight [19]. Oil tanker shipwrecks have caused a number of environmental disasters.

However, the number of oil tanker accidents has reduced, with oil spills of more than 700 tonnes reducing from 24 per year in 1970–79 to 7 per year between 1990 and 1999 [20]. But the ones that do occur result in major media coverage, and many still remember the wreck of the *Amoco Cadiz* in Brittany in 1978, which spilled 220,000 tonnes of crude oil and polluted 300 km of Atlantic coastline. That was one of the four biggest tanker accidents worldwide. The wrecking of the *Erika* on 12 December 1999 and the *Prestige* on 19 November 2002 are more recent. The *Erika* was a small ship that spilled 30,000 tonnes of

---

[4] The notation C_n indicates that the molecule contains n atoms of carbon ($C_1 = CH_4$, for example; C_4H_{10} is a molecule in C_4).

heavy fuel oil. But it should be remembered that illegal tank washing and degassing are responsible for 8–10 times more pollution than tanker shipwrecks.

2.3 Gas

Gas is formed at the same time as oil, but it migrates within the earth's crust more readily. Its use is relatively recent [21]. It is known that gas was used in China for cooking from the tenth century. This was firedamp conducted through bamboo tubes. In Europe, gas was more a curiosity at first, although some scientists experimented with the use of gas distilled from coal under pressure for lighting and inflating hot air balloons. The natural gas industry was born in the United States in the nineteenth century, the first exploitation dating from 1821. The first gas pipeline, 40 km long, was laid in 1870 using hollow pine trunks. Two years later, metal pipes were used to distribute gas to private residences.

World consumption of gas in 2000 was 2.164 GTOE, which represented 24.7% of global energy consumption [5]. French consumption of natural gas increased considerably after 1960 as can be seen from Figure 2.1. It increased from 2.5 MTOE in 1960 to 37.3 MTOE in 2000, a 15-fold increase.

More than 95% of gas consumed in France is imported.[5] The French gas field of Lacq, discovered in 1951, is now almost exhausted. The main gas suppliers to France are Algeria, Norway and Russia. The latter country passes 32.1% of world gas reserves [22].

Gas comes from on-land or undersea gas fields. The biggest gas field in the world is the Troll field [23], in the Norwegian sector of the North Sea. In 1999, this field provided 15 billion m^3 of the 38 billion that were consumed in France.[6] A submarine pipeline 840 km long connects the Sleipner and Troll gas fields of Norway to the French gas terminal in Dunkirk. Gas can be pumped down a pipeline at a speed of 30 km/h [23]. Compression stations were installed every 80–120 km to maintain this flow rate. This gas pipeline transports up to 14 billion m^3 of gas per year, or just over a third of total French consumption [25]. The total length of the transport and distribution network in France is close to 200,000 km.

Gas can also be transported in liquid form at a temperature of $-160\,^{\circ}\text{C}$. The volume of liquefied gas is 600 times smaller than the volume of gas vapour. Around 20% of gas worldwide is transported in this form [23]. A liquid natural gas (LNG) carrier 280 m long can carry some 130,000 m^3 of LNG. The energy contained in it represents more than 40 times that liberated by the explosion of the atom bomb over Hiroshima in 1945. Sea transport of gas is commercially viable for distances over 3000 km.

---

[5] In 2000, 4.28% of the 37.3 MTOE of gas consumed in France was produced locally.
[6] The conversion coefficient generally used is 1000 m^3 of natural gas = 0.857 TOE [5]. However, the calorific value of gas depends on its provenance, ranging for example from 33 MJ/m^3 for Dutch gas to 42 MJ/m^3 for Algerian gas [24].

Gas can be stored underground at pressures between 40 and 270 bar.[7] In France, there are 15 storage locations that make it easier to balance imported gas and consumption. There are 12 aquifer and 3 saline cavities representing about a third of annual consumption [24].

Gas use is still being expanded and this trend is likely to continue, for two reasons. First, gas is cheap for the consumer, even though its price follows, with a time lag, the price of oil; and second, the 'combined cycle' technology, which enables the energy contained in the exhaust gases of a combustion turbine to produce water vapour and use it in a steam turbine, enabling much higher energy efficiency (55%). The cost of a kWh then becomes very competitive. Also, the construction of a gas-fired power station is quick and needs a lower level of investment (460 euros/kW_e [26]) than a nuclear or coal-fired power station. The price of gas, however, represents 70% of the price of the kWh generated and there remain the problems of supply and price stability.

2.4 Greenhouse effect

All fossil fuels produce carbon dioxide when burnt and contribute to increase the greenhouse effect and cause climate change. However, the quantity given off depends on the type of fuel. The higher the content of hydrogen, the more carbon dioxide is given off. For the equivalent quantity of energy produced, coal, composed essentially of pure carbon, produces the most.[8] Gas, which is mainly made up of methane (CH_4), is the fossil fuel with the lowest emission. Oil, approximately $(CH_2)_x$, emits intermediate quantities. Table 2.1 shows carbon dioxide emission brackets for the production of 1 kWh of electricity.

Table 2.1 Carbon dioxide emissions per electrical kWh produced by different sources of energy

Method of production	Emissions (g/kWh)
Coal	860–1290
Oil	700–800
Gas	480–780
Nuclear	4–18
Wind	11–75
Solar photovoltaic	30–280
Biomass	0–116

The complete production chain is taken into account and not simply emissions during operation.

---

[7] In porous rocks (aquifers) or cavities excavated in salt deposits (saline cavities).
[8] There is approximately 1 atom of hydrogen for each atom of carbon in coal, or less than 10% in mass.

For the same fossil fuel, emissions depend strongly on the technology used. Gas with combined cycle turbines gives the lowest values. It is important to note that the values in the table apply to electricity generation. For simple heating, efficiencies are much better and emissions are lower. In addition, if heat and electricity are produced together (co-generation), the amounts of carbon dioxide produced per unit of energy generated are further reduced.

Table 2.1 also shows the emission brackets for other sources of energy. These figures refer, as for fossil fuels, to the complete energy production chain. Thus, although nuclear and photovoltaic power generators do not produce carbon dioxide during operation, greenhouse gases are emitted during the construction of nuclear power stations and photovoltaic panels and related transport operations.

2.5 Fossil fuel reserves

Nature stored fossil fuels for hundred of millions of years before human beings discovered them and began using them to meet their energy needs. As there is a finite quantity of these reserves on earth, it is pertinent to ask how long they will last at the rate we are currently consuming them. Since the industrial revolution, we have already used a large quantity of fossil fuels to ensure our economic development. The figures that follow should be considered as estimates subject to change with the discovery of new deposits and technological progress. They also need to be seen in the perspective of the 5 billion years that remain before our planet is swallowed up by the Sun.

2.5.1   Oil reserves

By 1997, rather more than 110 billion tonnes (Gt) of oil had already been extracted from underground. World reserves of cheap crude oil were evaluated in 1998 as about 120 Gt with a further probable 20 Gt to be discovered [27], in total 140 Gt. Since 1998, we have consumed oil at the rate of 3.5 Gt/year. Production in 2006 was 3.9 Gt [88]. Proven reserves of cheap oil were estimated by BP as 164.5 Gt at the end of 2006 [22]. Different estimates produce slightly different figures, but reserves are of the same order of size.

The idea of reserves is linked to the price that the consumer is willing to pay for access to the energy source. So, when the consumer price rises, some resources that were previously uneconomic become viable and increase the level of reserves. Price rises can also cause a drop in consumption, and vice versa. Three different quantities are assessed in evaluating oil reserves [27]. The first is the quantity of oil already extracted. This is the value where there is the least uncertainty. The second is the quantity of extra oil that could be extracted from existing wells through advances in technology. Also, some known deposits have not yet been exploited. The third quantity is an estimate of the oil that there are reasonable chances of discovering and exploiting. These estimates are weighted with a larger or smaller probability factor according to one's optimism or pessimism. Non-scientific factors may distort the estimates, because some countries have an interest in overestimating their reserves in order to export more or ask for loans.

Around 80% of oil extracted today comes from deposits discovered before 1973. Experts believe that 90% of oil reserves have already been discovered and that the production of cheap oil, that is what we are currently consuming, will decline in the next 10 years. Their arguments are based on an extension of the model developed by King Hubbert in 1956, which predicted that the uncontrolled extraction of a finite resource follows a bell-shaped curve the peak of which occurs when half the resource has been exhausted (Figure 2.2). When applied to the United States in 1956, this model predicted that the oil production of the country would peak in 1969. This peak was actually reached in 1970 and production followed the bell-shaped curve fairly closely.

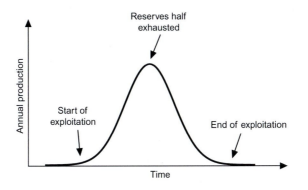

Figure 2.2   The Hubbert theory of the exploitation of a finite resource

Even if oil reserves were far in excess of 140 Gt, one would still expect a decline in production within 20 years. If this value was doubled (280 Gt), peak production would be reached around 2020. Cheap oil will therefore become rarer and, depending on global economic development, prices would increase within one or two decades. The world will become more and more dependent on the oil-producing countries of the Middle East (Saudi Arabia, Iran, Iraq, Kuwait and United Arab Emirates). This dependence may lead to price instability, as occurred in 1973 and 1979. In 1973, the OPEC countries increased their prices because they controlled 36% of the market. Prices only fell when demand was reduced and the North Sea and Alaska fields came on stream [27].

Table 2.2 shows the distribution of oil reserves by world region. As can be seen, the largest reserves are in the Middle East. With around 164.5 Gt and a current production of some 3.9 Gt,[9] the reserves will last about 40 years (164.5 [over] 3.9 ≈42 years). Most of these reserves are in the Middle East. Figure 2.3 shows the seven countries with the biggest oil reserves.

Table 2.2   Proven global oil reserves by region at the end of 2006

Region	Reserves (Gt)	%
North America	7.8	5
South and Central America	14.8	8
Europe and Eurasia	19.7	12
Middle East	101.2	61.5
Africa	15.5	9.7
Asia Pacific	5.4	3.4
Total	164.5	100

Totals do not exactly equal 100% because of rounding errors (Source: BP Statistical Review of World Energy, June 2007).

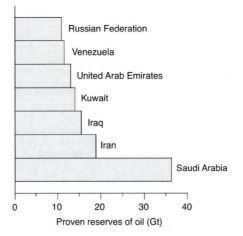

Figure 2.3   Countries with the largest proven reserves of cheap oil at the end of 2006 [Source: BP Statistical Review of World Energy, 2007]

In fact, these estimates do not mean that there will be no more oil in 40 years, but that it have become much rarer and more expensive. By 2050, world oil production is estimated to be of the same order of size as today, but requirements will have increased. For some 20 years now, less new conventional oil has been discovered than is consumed.

[9] In 1991, 300 Mt of oil were burnt in Kuwait during the first Iraq conflict, or more than three times French oil consumption in that year.

As well as crude oil, there are other fossil petroleum resources, known as unconventional, which could replace oil in the future. Some are already exploited on a small scale. These include heavy and ultra-heavy fuel oils, bitumens, bituminous shales and tar sands.

Heavy oils and bitumen arise from the transformation caused by bacteria that consume the lighter fractions of the oil. The viscosity of heavy crude oil is 60 times that of conventional crude oil. Bitumen is even less viscous. Reserves are estimated at 90 GTOE for heavy oils and 100 GTOE for extra-heavy and bitumen [28]. Canada (Athabasca region) and Venezuela (Orinoco Belt) each have reserves amounting to 48 GTOE, or 58% of known global reserves (190 Gt) of these fuels. A number of methods can be used to extract these substances but the most usual one is by steam, which needs energy to produce it. Another possibility might be to use small nuclear reactors locally to produce heat for the steam extraction process instead of burning fossil fuels.

Bituminous shales are another source of oil. These are sedimentary rocks containing kerogen, the precursor of oil and gas, which can be transformed underground into oil and gas by cracking at high temperature. Shale is called bituminous if it contains at least 40 l of equivalent oil per tonne. Shales have been known and exploited for 150 years but are rarely economically viable. To be economic they must contain at least 80 litres of oil per tonne. They are used to produce electricity (17–18 Mt/year) or oil (720 TOE/year) [28]. Bituminous shales are found in many parts of the world, but 50% of reserves are in the United States. Proven reserves amount to 70 GTOE, with additional reserves of 60 GTOE, a total of 130 GTOE [28]. Considerable uncertainty exists on the global level of existing reserves. One estimate is between 510 and 540 GTOE. Add to this a low extraction rate of the order of 10%, which considerably reduces the quantities that could really be used. Existing reserves must not be confused with exploitable reserves that are very often much lower.

The extraction of heavy crude oil, bitumen, bituminous shale, etc., demands much more energy than the extraction of conventional oil, which adds to the cost and associated pollution. Also, given the uncertainties, the values given should be taken as approximations.

2.5.2 Gas reserves

Gas reserves are abundant and every year more gas is discovered than is consumed [25]. Proven reserves (whose existence has been physically demonstrated) are around 180,460 billion m^3 or 162.4 GTOE[10] [22]. Ultimate reserves, which are arrived at by a statistical evaluation of the quantities judged to be recoverable using existing methods of exploitation, are around 400,000 billion m^3. In 2006, commercial production amounted to 2865 billion m^3 or 2.6 GTOE, representing around 24% of primary energy consumption. At this rate of extraction, there are 63 years of reserves – more than oil (40 years). However, higher demand would

---

[10] Using the equivalence of 1 Gm^3 = 0.9 MTOE.

reduce this time frame. Thus, if one were to replace oil and coal by gas to reduce the greenhouse effect, there would be less than 20 years of reserves.

Table 2.3 shows the distribution of world gas reserves. The largest reserves are in the Russian Federation (26.3%), Iran (15.5%) and Qatar (14%). Figure 2.4 shows the countries possessing the largest proven gas reserves.

Table 2.3   World gas reserves by region

Region	Reserves (10^{12} m^3)	%
North America	8	4.4
South and Central America	6.9	3.8
Europe and Eurasia	64.1	35.3
Middle East	73.5	40.5
Africa	14.2	7.8
Asia Pacific	14.8	8.2
Total	181.5	100

(Source: BP Statistical Review of World Energy, June 2007).

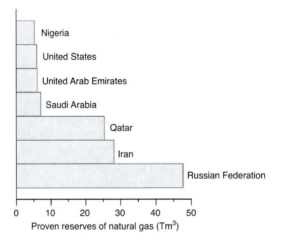

Figure 2.4   Countries with the largest proven gas reserves at the end of 2006 [Source: BP Statistical Review of World Energy, 2007]

Enormous quantities of methane are stored in the form of *methane hydrate* in the permafrost (permanently frozen soil), or at great depths under the sea. Methane hydrates are complexes (clathrates) in which a molecule of gas is trapped within molecules of water. They are formed under pressure. Oilmen are familiar with them because they can appear in offshore oil pipelines and block them. A given volume of water can store as much as 160 times its volume of methane. Estimates of methane hydrate reserves vary widely between 1 and 5 Pm^3 (1 Pm^3 = 1015 m^3), which is almost as much as oil, coal and gas reserves combined [89]. At present there is no way of exploiting this resource. No doubt it will be possible one day, but the process would present environmental dangers because of the methane that could escape into the atmosphere and that has a far more toxic greenhouse effect than carbon dioxide. Climate change caused by fossil fuel combustion could also, through associated global warming, liberate part of the methane trapped in the hydrates and powerfully amplify the greenhouse effect.

2.5.3 Coal reserves

Coal is the fossil fuel with the largest reserves, estimated to be 518.9 MTOE[11] in 2002 [5]. At the end of 2006, total reserves were as follows: 478.8 Gt for coal classified as anthracite and bituminous according to the American classification, and 430.3 Gt for sub-bituminous coal and lignite.[12] For both categories combined, the largest reserves are found in the United States (27.1%), Russian Federation (17.3%), China (12.6%) and India (10.2%). For good quality coal (anthracite and bituminous – *hard coal*), the three countries with the biggest reserves are the United States (111.3 Gt), India (90 Gt) and China (62.2 Gt). For poor quality coals (sub-bituminous and lignite – *brown coal*), the three countries with the largest reserves are the United States (135.5 Gt), Russian Federation (107.9 Gt) and China (52.3 Gt). Table 2.4 [20] summarises the data for the different regions of the world.

Table 2.4   World coal reserves by region and type of coal

Region	Anthracite and bituminous (Gt)	Sub-bituminous and lignite (Gt)	Total reserves (Gt)	%
North America	115.7	138.8	254.4	28
South and Central America	7.7	12.2	19.9	2.2
Europe and Eurasia	112.3	174.8	287.1	31.6
Middle East and Africa	50.6	0.2	50.8	5.6
Asia Pacific	192.6	104.3	296.9	32.7
Total	478.8	430.3	909.1	100

Totals do not exactly equal 100% because of rounding errors (Source: BP Statistical Review of World Energy, 2007).

---

[11] The conversion factors used for this estimate are 0.7 TOE/t for bituminous coal, 0.5 TOE/t for sub-bituminous and 0.3 TOE/t for lignite.

[12] There are several classification systems for coal – two French and one American. According to the latter, coal is classified by its calorific value that leads to a distinction between *hard coal* (anthracite and bituminous coal) and *brown coal* (good and bad lignites, sub-bituminous coals).

Combining the totals for the *hard* and *brown coal* gives 909.1 billion tonnes, which corresponds, at the current rate of production (3.1 GTOE in 2006), to sufficient reserves for about two centuries [22]. Figure 2.5 shows countries with the biggest reserves (hard and brown coal combined).

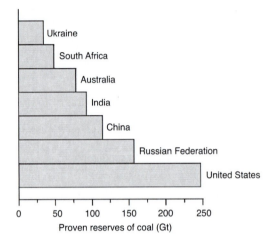

Figure 2.5   Countries with the largest proven coal reserves at the end of 2006, including all types of coal (anthracite, bituminous and subbituminous, lignite) [Source: BP Statistical Review of World Energy, 2007]

Coal is a bedrock for methane that often remains trapped in the layer where it was created in the form of a film around particles of coal. In a given space, its concentration can be 6–7 times higher than that found in a gas field. It is estimated that some 200 m^3 of methane may be formed during the creation of 1 tonne of coal. The quantity of methane, known as 'coal gas', trapped in coal is grossly estimated at between 93 and 185 GTOE (100–200 Tm^3) [28]. The rate of recovery of this resource is low, around 14% in the United States, which gives reserves of between 15 and 25 GTOE.

2.6   Conclusion

Because of the convenience and high energy density, fossil fuels have consistently proved superior to renewable energy sources. Coal was the energy of choice from the middle of the nineteenth century and oil from the beginning of the twentieth century. Despite their advantages, their widespread use has had a measurable impact on the environment, increasing the greenhouse effect and causing various kinds of pollution. Fossil fuels are also present in finite quantities on earth that raises the problem of their increasing scarcity.

Coal is the most polluting, but its reserves are larger than those of any other fossil fuel. Oil is the most convenient fuel, being liquid and easily transportable. It is still irreplaceable for motorised transport. The main advantages of gas are its low level of pollution and the high efficiency of its combustion. One problem is leakage during its distribution and use because methane is 23 times as potent a greenhouse gas as carbon dioxide.

Technology makes it possible to convert one type of fuel to another, for instance coal to oil or gas. This reduces the differences existing between different types of fossil fuels and will be useful when oil and gas become more scarce.

Technology for synthesising liquid hydrocarbons from coal is a very important challenge for the future. At present, oil is irreplaceable for motorised transport. Synthetic fuels produced from coal could be far more useful than hydrogen and considerably extend the useful life of the combustion engine who's technology and cost are well understood. A coupling of the internal combustion engine with a system made up of batteries and an electric motor (hybrid vehicle) makes it possible to reduce local pollution and emit less greenhouse gas. The use of hybrid vehicles is expanding rapidly and already one million Toyota Prius have been sold since the launch of this model.

Gas consumption is expanding rapidly. This source of energy calls however for heavy investment and infrastructure costs. Also, the cost of transport is an important part of the price of the final kWh.

Oil's widespread use arises from its convenience. As a concentrated liquid source of energy easily and cheaply available, it has made economic development dependent on it. Unlike natural gas, the cost of transport only represents 5–10% of its price (it only costs slightly more than one euro to transport one barrel of oil from the Middle East to Europe). More than 60% of oil is shipped by sea, the remainder being transported by pipeline, train or trucks. In 1920, world production of oil was only 95 MTOE whereas today it is around 3500 MTOE. There will still be oil at the end of this century and perhaps beyond. However, its price will rise and it is essential to use it sparingly and keep it for uses where it is indispensable – chemical and transport industries. Synthetic oil could also be produced from coal (CTL = coal to liquid), from gas (GTL = gas to liquid) or biomass (BTL = biomass to liquid).

In 2000, fossil fuels represented nearly 90% of global primary commercial energy consumption. Their supremacy has not arisen by chance. It comes from their flexibility in use and their economic competitiveness. They will remain the main sources of energy for a long time to come, but this should not prevent us thinking now about substitutions for the day when they will run out, even if that is not going to happen tomorrow.

Chapter 3

Renewable energies

Renewable energies will be available as long as the Earth exists – another 5 billion years. They are sometimes referred to as *new energies*, although they have been known for many centuries. They were used by primitive humans who, after discovering how to make fire, burnt wood to warm themselves and cook their food. Renewable energies have been used for a longer period than fossil fuels or nuclear energy. They had the monopoly of world energy production until the end of the eighteenth century and still met half of human's energy needs at the beginning of the twentieth century. Today, renewable energies have been rediscovered thanks to technological progress that has been able to remove their main drawbacks – low intensity and intermittent nature.

There is a tendency at present to associate renewable energies simply to the generation of electricity. In fact, they make a major contribution to heat production as can be seen in Figure 3.1, which gives the distribution of renewable energy output for 2005 in France. Heat production represents 67.4% of the use of renewable energies.

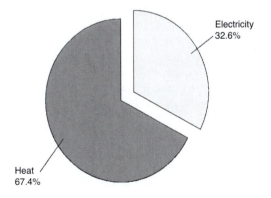

Figure 3.1 *Distribution of renewable energies in France in 2005 between electricity generation and heat production. The conversion between electricity and heat is based on energy content (1 MWh = 0.086 TOE). [Source: [3]]*

Renewable energies are free, but their exploitation can be expensive, even very expensive, because they are not highly concentrated. While they are well suited to a habitat with low population density and low energy demand, their diffuse potential is incompatible with the spatial concentration of urban and industrial energy demands. Hydropower is a special because it can produce considerable amounts of energy. For example, the Three Gorges Dam in China has an electrical generating capacity equal to 18 nuclear power stations or 1 GW_e.[1] But apart from hydropower and some specific situations, the cost of electricity produced by renewable energies is considerably higher than that of standard techniques based on fossil fuels or nuclear power, except in very particular situations. The lack of economic competitiveness and their inter-mittent nature are a brake on their development. This explains the financial and fiscal incentives introduced by many countries to encourage their uptake.

3.1 Hydropower

The motive power of water was one of the first energies used by human, who harnessed it to mill grain, saw wood, and work forge bellows and weavers' looms. The watermill was known in Syria before the Christian era and was also used by the Romans. The introduction of the cam in the eleventh century enabled watermills to be used as sources of mechanical energy that were unsurpassed until the arrival of the steam engine. Until the end of the eighteenth century there was little technological progress and most watermills functioned with wheels with horizontal vanes. Their efficiency increased noticeably in the nineteenth century, thanks mainly to the work of Poncelet. The introduction of the steam engine gradually caused watermills to disappear, but energy generated from the move-ment of water came back with the introduction of electricity.

A quarter of the energy that comes from the sun is absorbed by the water cycle. In the process, a tiny part of the water of the oceans is evaporated by solar energy and returns to earth in the form of precipitations of rain or snow. Part of this falls on the continents, which only represent 30% of the earth's surface, before being evaporated or rejoining the sea. This contributes to the growth of plants and results in the formation of streams and rivers.

Like all renewable energies, water energy diffuses. The terrain, small streams, torrents and rivers, all help to concentrate the rain into lakes that enable energy to be stored.

The exploitation of hydropower makes use of the difference between the potential energy of water according to its height above sea level. A body of mass m situated at height h has, in the earth's gravitational field, whose acceleration is g, an energy potential of mgh. If it falls from height h_1 to h_2, where $h_1 < h_2$, its energy reduces by $mg (h_1 - h_2)$. Part of this energy can be recovered when the

---

[1] The Three Gorges Dam will be more than 2 km wide and its reservoir more than 640 km long. The depth of the basin will be 185 m.

water is directed into a turbine. The energy that can be extracted from a waterfall is proportionate to its rate of flow and height. There are high falls with a low rate of flow, and low falls with a high rate of flow. Large quantities of water and a substantial change of level are necessary to produce large quantities of electricity, for example, 1 kWh represents the energy of 3.6 tonnes of water falling from a height of 100 m. A family refrigerator consumes on average 380 kWh/year. To run a refrigerator with electricity from hydropower would take the energy of nearly 1400 tonnes of water falling from a height of 100 m, which is equivalent to 3.8 tonnes of water per day falling from this height.

A dam is used to retain water in a natural or artificial lake or sometimes a watercourse, by transferring the weight of the water onto the bottom or lateral walls. There are two types of dams; the first type uses larger masses of material than the second, but the second type requires very strong lateral walls to resist the weight of water. To produce electricity, the water is channelled into a turbine that generates mechanical energy. This is coupled to an alternator, producing electricity. The various types of turbine have replaced the watermill because the efficiency is much higher (from 70% to 200% instead of 20%).

The exploitation of hydropower is closely linked to the flow rate of watercourses. These vary according to the region and season and some years are better than others. For instance, in tropical countries it rains only in the summer, whereas in temperate climates there is rain almost throughout the year. Reservoirs help to cancel out this intermittence and allow large quantities of energy to be stored, which can then be used to meet local demand.

Theoretically, the global hydropower potential is 36,000 TWh, but in fact only 16,000 TWh is exploitable and of those only 8000 TWh is economically viable. Currently, less than 20% of this potential is used. Hydropower is the most widely used commercial source of renewable energy, providing currently between 13% and 14% of electricity in countries of the OECD. However, in Europe, further extension of hydropower is difficult because it has already been fairly fully exploited. In Asia, Latin America and the former USSR, on the other hand, hydropower is under-exploited. In France, 56% of all electricity production was produced by hydropower, in 1960.

Figure 3.2 shows the production of hydropower in the main producer countries in 2004 [91], and compares the figures for the EU, former USSR and France, which comes out well considering its geographic area. World production of hydropower in 2004 was 2889 TWh, 16.5% of total energy production (17,525 TWh).

Industrial hydro installations generate several MW or even in excess of 10 GW for the larger dams, while small-scale hydro plants produce outputs below 10 MW: between 2 and 10 MW for small, from 0.5 to 2 MW for mini- and below 0.5 MW for micro-hydro stations [30]. *Damless hydro-plants* capture the kinetic energy of rivers and other waterways without using a reservoir, but other energy sources are needed for isolated areas during the period when there is not enough water.

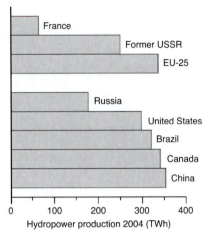

Figure 3.2   Major producer countries of hydropower in 2004 (bottom), and, for comparison, the production of the EU 25, former USSR and France [Source: [91]]

Hydropower plants are expensive to install but economic to operate. In France, the investment for hydroelectric dams was hardly less than the cost of the nuclear power stations. Industrial hydro is particularly economical, with a kWh costing around 2 euro cents. Small-scale hydro is rather more expensive, around 4 cents. But the potential of small-scale hydro is insufficiently exploited in France and much more could be installed if the problems of acceptability and environmental impact can be overcome.

Huge amounts of energy could be recovered from the oceans, but its recovery presents considerable difficulties. The best known system is the tidal barrage, which has the advantage of the timing and intensity of the tides being known in advance. Tides result from the gravitational pull of the sun and the moon on the oceans. Their size depends on the relief of the seabed. In a tidal power station, a basin that is filled by the sea at high tide and retained at low tide by a dam is created. Driving a turbine, energy is recovered by water from the basin. It is proportional to the surface of the basin and to the square of the size of the tide, which varies by a factor of 2 between high and low tides, hence a factor of 4 in energy. The tidal power station of the Rance in Brittany (240 MW) is the largest in the world, producing around 0.5 TWh/year. This relatively low output arises from the fact that it is in operation for three times less than a traditional power station in a year. The cost of the electricity generated is high, between 5 and 11 euro cents per kWh. The exploitable potential of tidal power generation in Europe is 105 TWh, mostly in France and the United Kingdom.

There are other forms of marine energy apart from tidal [90]. They are not yet economically viable but may prove worthwhile in the future. They include offshore wind generation, wave generation, marine current generation, thermal energy and osmotic energy.

Offshore wind generation is more expensive than land-based wind generation, but this is partly compensated for by the better wind quality at sea. The potential of this resource is 1800 TWh/year. To add to the problems of land-based wind generation are those of fixing the generators on the seabed and of salt water corrosion. Offshore wind generators currently have rotor diameters of over 100 m and generate from 2 to 5 MW. They operate at full power for more than 3300 h/year. The density of power per surface unit is around 6 MW/km^2, which enables some 20 GWh/km^2 to be produced. This is 15 times less than photovoltaic panels for the same surface, but in the case of wind farms the surface is only partially occupied by the turbines. Investment costs around 1500–2500 €/kW or even more in difficult sites.

France has good wave energy resources. Global resources of this type of energy are estimated at 1400 TWh/year. In the case of France, the average energy transmitted is around 45 kW/m. The cost of the electricity generated varies between 0.05 and 0.1 €/kW for an investment of between 1000 and 3000 €/kW. One of the problems of this form of generation is the likelihood of damage by storms.

Marine currents, like the wind, transport kinetic energy, and this energy can be partially recovered by underwater rotors similar to wind rotors. As the density of water is 837 times that of air, marine current rotors can be smaller, even though the speed of currents is much lower than that of wind. For a given size, a marine current generator is around four times as efficient as a wind generator. For a current of 3 m/s, energy of 4 kW/m^2 can be generated. Marine generators can easily operate for 4000 h/year and produce electricity at a cost of 0.06–0.08 €/kWh.

The thermal energy of the ocean is of interest in hot countries, and estimates have placed its potential around 80,000 TWh/year. The idea is to exploit the temperature difference between the surface water and the water at great depths, typically around 20–24 °C, through a heat exchanger. The output is low, between 2% and 3%. At present, this is a very expensive form of energy.

Osmotic energy makes use of the difference in salt concentration between sea water and freshwater by means of a semi-permeable membrane. The freshwater migrates through the membrane to the sea water. The difference in osmotic pressure is 27 bar for sea water but the operation is carried out at an over-pressure of 10 bar. Large membrane surfaces are required (between 200 and 250 m^2/kW).

The potential of marine energies is considerable. Their cost is still too high, but they have the advantage of producing electricity without the emission of greenhouse gas. Their exploitation will never be easy, and will have to be acceptable to other users of the sea, and be resistant to extreme weather conditions and corrosion. A further problem is that the energy produced then has to be transported to dry land.

3.2   Solar energy

The luminous flux reaching the Earth from the Sun is 1368 W/m^2. This is the total quantity of energy received by surface unit and time from outside the Earth's atmosphere, but the actual amount of energy arriving at ground level is less. The highest energy achieved is around 1000 W/m^2 (when the Sun is at its zenith). At the Earth's surface, not all wavelengths of the Sun's radiation are received at the same strength, and fortunately dangerous ultraviolet radiation is filtered by the ozone layer.

There are many reasons why solar radiation is modulated at the Earth's surface: day and night, the seasons, the latitude and cloud cover. Averaging the light energy received over a year results in a figure of some 20% of the solar luminous flux and often less [31]. It varies between ∼6 kWh/m^2/day in deserts like the Sahara and <1 kWh/m^2/day in the Polar Regions. In France, solar radiation varies between 3 (Lille in the North) and 5 kWh/m^2/day (Nice in the South). For countries in the Tropics, the major proportion of the energy received comes from direct solar radiation. In the countries of Northern Europe, more than half of the energy received comes from diffused radiation.

The solar energy arriving every year on the Earth is equivalent to 1.56×10^{18} kWh, or around 15,000 times the current global energy consumption. Of this energy, 30% is reflected back into space, 45% is absorbed, converted into heat and returned to space in the form of infrared radiation, while the remaining (25%) contributes to evaporation (22%), wind kinetic energy (2%) and photosynthesis (0.06%).

Solar energy can be used to produce both heat and electricity – *solar thermal* and *photovoltaic* energy.

3.2.1   Solar thermal

Everyone enjoys the warmth of the Sun's rays after a hard winter. This energy, which we feel in the form of heat, can be harnessed whenever the Sun shines. It is used passively in well-designed buildings that make intelligent use of the solar warming effect and heat loss. Openings in the structure, the type of glazing and insulation all play an important part in the energy balance of a building. The use of appropriate materials enables heat to be stored when there is sunshine, to be reused when there is none. Architecture can also play a significant role in maintaining a constant temperature in buildings. This passive use of solar energy contributed nearly 10% of the total energy used in the European Union in 1990 [30]. In hot countries, good architecture can reduce or even eliminate the need for air conditioning.

The simplest application of solar energy is to use it for heating a fluid from which heat can be extracted directly or indirectly through a heat exchanger. When solar energy has raised the temperature of this fluid to 50 °C or 60 °C, it can be used to heat domestic hot water or radiators. With under-floor heating, the fluid temperature can be as low as 25 °C. The heating elements are metallic panels with a blackened surface to better absorb the solar radiation. A small pump is used to transfer the heated fluid to a heat exchanger.

Solar domestic hot water heaters are the simplest solar energy appliances. The fluid heated initially may be water that is used directly if there is no danger of frost in the winter, or glycol used with a heat exchanger to heat a hot water cylinder. Extra heating by a boiler is necessary to raise the temperature in the absence of solar radiation. Solar collectors produce between 200 and 800 $kWh/m^2/year$ [90]. A solar panel with a surface area of 3–5 m^2 is sufficient to provide hot water for a family of four [30] and represents an average investment of between €2000 and €4000. Solar collectors can also be used for heating a dwelling (or a swimming pool) by under-floor heating. Approximately 1 m^2 of collectors are needed per 10 m^2 of dwelling and to provide some of the heating requirement by under-floor heating it takes between 10 and 20 m^2 of collectors, which would save 50–60% of heating costs. The thermal efficiency of solar collectors is around 50%.

Solar thermal energy can also be used to produce electricity through the thermodynamic process. The high temperatures necessary for this are achieved by concentrating sunlight with mirrors. This enables temperatures of several hundred degrees and sometimes of over 1000 °C to be attained. Power stations with parabolic reflectors are already in service (354 MW installed in California [30]). Other types of solar power stations are so far only experimental, such as the Themis solar tower in south-western France. Efficiency is around 10–15% and the price of the kWh of electricity generated is between 9 and 15 euro cents – too expensive compared to standard methods of production. They are more worthwhile in hot countries where electricity is needed for air conditioning when the sunshine is hottest.

A concrete solar tower 1 km high, surrounded by 38 km^2 of solar thermal panels, is currently in the planning stage in Australia. The air heated by the solar panels is directed towards the 130-m-diameter tower, creating a draught moving at between 35 and 50 km/h. The temperature at the top of the tower is much lower, because the temperature falls on average by 1 degree for every 100 m of altitude. The air current drives 32 turbines situated at the base of the tower. This solar power station should deliver 200 MW, or one-fifth that of a standard nuclear power station. It will work day and night.

3.2.2   Photovoltaic

The photovoltaic (PV) effect was demonstrated by Becquerel in 1839 by illuminating an electrode submerged in an electrolytic liquid. After the discovery of the photoconductivity of silicon by Smith in 1873, the same effect was observed on a solid material by Adams and Day in 1877. The first PV cell, constructed in 1914, had an efficiency of 1%; it was used to develop exposure meters for photography. The PV production of electricity was not developed until 1954, with the manufacture of the first monocrystalline silicon cell, which had an efficiency of 6%. Efficiency soon improved, however, to reach 14% in laboratory conditions in 1958.

PV applications began in the early 1960s with their use in satellites, then in consumer goods such as watches and calculators, and in electricity production systems for remote sites.

The cost of PV electricity is much higher (normally by a factor of about 10) than that obtained from fossil fuels or nuclear, but this price is falling. PV is a very useful source of energy for remote areas because it saves the need to invest in kilometres of electric cable (1 km of line costs between €15,000 and 30,000).

PV is a decentralised energy source, useful in sunny areas for relatively low energy requirements. It can provide some lighting, drive a water pump and run a refrigerator to preserve drugs and vaccines – or any equipment that needs only around 1 kWh/day.

The conversion of light energy to electrical energy is accomplished with the use of semiconductor materials. The PV market is dominated by mono-crystalline, polycrystalline or amorphous silicon. Monocrystalline silicon PV cells are obtained by making very fine slices of 300–400 μm from a block of monocrystalline silicon manufactured by the slow cooling of molten silicon.

Polycrystalline silicon is obtained from a silicon block cooled more rapidly. It consists of multiple small silicon crystals separated by a visible grain. The manufacture of a rod of polycrystalline silicon is quicker and requires less energy than a monocrystalline rod, but the cells obtained have a slightly lower conversion yield.

Unlike the two previous types of silicon, amorphous silicon is a non-crystalline allotropic form of silicon. Its advantage is that it can be continuously deposited in a thin film on a glass substrate, but its PV yield is lower.

In a semiconductor material, the concentration of charge carriers is much lower than in metals. The energy bands are characterised by the existence of a 'forbidden band', separating the valence band from the conduction band. To excite an electron to pass from the valence band to the conduction band, a minimum amount of energy called 'band gap' has to be supplied. This is a specific characteristic of semiconductors. This band gap energy is 1.1 eV for silicon and 1.7 eV for amorphous silicon. When photons from the Sun strike the cell, some are reflected, some pass through the cell and some are absorbed. The absorbed photons whose energy is superior to the semiconductor band gap can create an electron–hole pair (an electron in the conduction band and a hole (empty space left by the missing electron) in the valence band). To obtain an electric current, the electron and the hole have to be separated, by creating an electrical field in a semiconductor assembly, a diode *p-n*, for example. Zone *p* corresponds to the semiconductor doped with electron-accepting atoms in a crystalline lattice, which creates excess holes. Zone *n* corresponds to the semiconductor doped with electron-donor atoms, which gives excess electrons. The excess electrons in zone *n* and the excess holes in zone *p* created an electrical field that enabled the charges generated by the photoelectric effect to be separated. With silicon, a zone *p* can be created by doping it with boron and a zone *n* by doping with phosphorus.

The efficiency of industrially manufactured PV solar panels varies from 13% to 18% for crystalline silicon and 7% to 8% for amorphous silicon, the efficiency of which reduces to 6% during the first weeks of use before it sta-bilises.[2] Until recently, silicon scrap from the microelectronic industry was remelted and then sliced, but the quantities are now insufficient. Few busi-nesses are prepared to invest in the manufacture of solar quality silicon, but the industry is coming up with short-term solutions that use cheap material – rib-bons, thin film, Smart Cut – that are now starting to appear.

A PV cell delivers a direct electric current of < 1 V with a power of around 1–3 W. Commonly used voltages are 12, 24 and 48 V, which require several cells to be connected in series. To obtain a higher power, these cell series are linked in parallel, enabling panels with a power of several tens of watt-peak (W_p).[3] It should be noted that the power delivered by a panel falls as the temperature increases.

Most PV modules are made of slices of silicon, doped for the reasons explained above, surface treated with titanium oxide for example, then encapsulated. A PV generator is made by connecting a group of modules. A PV system comprises a rigid structure to support the modules, cables, a battery and its charge regulator if electrical energy is being stored and an alternator when feeding equipment using alternating current.

Let us look for a moment at the capacity of electricity production from PV cells. The power received in full sunlight is 1000 W/m^2. The average efficiency of a cell is around 10%, because it falls as the temperature rises. The equivalent of a day's exposure at full power is of the order of 3–4 h/day, or about 1000–1500 h/year. This gives a daily production of 300–400 Wh/m^2/day, in other words between 100 and 150 kWh/m^2/year. A panel of 1 watt-peak generally provides 1 kWh/year. The world production of PV electricity was some 840 GWh in 2004 [90, 91], or almost 3500 times less than industrial hydro, for rather less than 1200 MW$_p$ installed. The three countries with the largest installed PV capacity are Germany, Spain and Italy, with 40% of the installed world PV power. [92].

The advantages of PV are its reliability, adaptability, low cost in use and the fact that operation is non-polluting. However, the manufacture of solar cells uses large quantities of energy[4] and emits polluting substances.

The cost of PV panels has fallen but there remains a long way to go. The price per watt-peak fell from €12–15 in the 1980s, to €6 in the 1990s and €3.30 in 2000. Because of high demand, the price of the raw material (electronic

---

[2] These efficiencies correspond to the industrial manufacture of panels of several m^2 in the laboratory (surfaces from 1 to 100 cm^2), or for short series, efficiency can be much higher. For example, for crystalline silicon, efficiencies of 26% can be achieved in the laboratory and 22% in short series.
[3] The power of solar panels is generally measured in watts-peak – the power delivered in reference conditions corresponding to a solar luminance of 1000 W/m^2 and a temperature of 25 °C.
[4] For example, the manufacture of a 1-W module in monocrystalline silicon requires some 5 kWh of energy. Four to five years of operation are necessary before the panel has recovered the energy that it took to manufacture it.

grade silicon) increased from $15/kg in 2000 to $75–100 in 2005 [90]. Demand has led to increased competition in the manufacture of quality solar material.

PV applications are booming worldwide. Global production of PV modules rose from 5 MW_p in 1982 to 60 MW_p in 1992. By 2000, it was 280 MW_p, and by 2001, 385 MW_p. The final cost per watt-peak is different according to whether the system is coupled to the electrical grid or isolated. In the first case, an alternator is needed, which leads to a price bracket between 3.80 and 5.30 €/W_p. In the second, a storage battery is needed, the cost of which varies between 1.50 and 4 €/W_p, which leads, for the whole system, to a cost of between 5.30 and 9.30 €/W_p.[5]

For PV connected to the grid, the price of the kWh is between 25 and 50 euro cents. That is not economical in a country like France, but could become so through the installation of roofs with integrated PV and thermal panels, which could generate heat and electricity as well as providing shelter.

An isolated PV system is more expensive because batteries must be included. The price of a kWh is between €0.75 and 1.50. It is useful for remote sites where a connection to the grid would be much more expensive. In developing countries, the price is competitive compared to the use of small generators, candles, kerosene lamps or dry batteries, whose cost can amount to hundreds of euros for 1 kWh.

One of the major problems for autonomous PV systems is the lead battery. While a solar panel can last for 30–40 years, batteries must be replaced. Good quality batteries (tubular batteries) may last as long as 12 years, but mediocre ones only last 2–4 years. Since a cheap lead battery must be replaced in 2 years if it is recycled and 4 years if it is not, it is clear that the energy needed to manufacture them will never be recovered.

3.3 Wind energy

Wind energy has been in use for a very long time, particularly for navigation by sail, but also by windmills to draw water from wells and grind grain. In the fifth century BC, the first windmills with vertical axes were being used as a source of mechanical energy. Around the same time, windmills with horizontal axes were in use in Egypt [32]. Similar windmills were developed in Europe from the seventh century and their technology became increasingly complex, culminating in the windmills of Holland. The concept of making electricity from wind power dates from 1802 (Lord Kelvin), but the development of modern wind generators did not begin until the invention of the dynamo in 1850. The system generating electricity from wind power is called an *aerogenerator*.

---

[5] The PV module and the batteries represent, respectively, 67% and 14% of the initial investment, whereas after 20 years the module only represent 33% of the cost and the batteries 48%. This is because of the long life of the module (more than 30 years), while the batteries have a much shorter life (generally 5 years) and have to be replaced several times.

Wind power was widely used (there were 300 aerogenerators in France in 1920) until the price of oil made this source of energy uncompetitive. But since the 1980s, there has been a sharp increase in wind generation. At the beginning of the 1980s, the total installed wind power connected to the grid was very low, then it rose to 2 GW by 1990, 13.4 GW in 1999, 16 GW in 2000, around 20 GW in 2001 [32] and 59.3 GW at the end of 2005 [90]. Most wind power generator is installed in Europe (4.5 GW) with Germany (18.4 GW), Spain (10 GW) and Denmark (3.1 GW) in the lead. Germany and Denmark are close to saturation and yet in Denmark wind energy only represent 15% of the total electricity production [90].

Wind energy is generated by the unequal heating of the Earth's surface by solar radiation, because more energy is absorbed at the equator than at the poles. This creates differences in heat and pressure that create air currents. At the planetary level, the wind currents flow from zones of high pressure to zones of low pressure. They are strongly influenced by the Coriolis force, which acts in a direction perpendicular to the direction of movement in the northern hemisphere and in the opposite direction in the southern hemisphere.

The ocean and the continents, which absorb solar radiation differently, also generate winds. The onshore and offshore winds known to sailors occur because of the difference in temperature of the air above the sea and the air above the land on the coast.[6]

Differences in temperature and air pressure created by solar radiation induce unpredictable instabilities that generate the winds. In any region, the wind speed varies over the course of time. Only observations over several years enable one to know statistically whether a particular site has sufficient wind and the main characteristics of wind during different times of the year.

The power per unit of surface of a fluid is described by $P = \rho V^3/2$ where ρ is the density of the fluid and V its hydrodynamic speed. It is not possible to recover the whole of this power, because the wind speed behind the blades of a wind generator or windmill can never be zero. It can be shown that, for an incompressible fluid, the maximum theoretical efficiency is 16/27 ($\sim$60%) of the kinetic energy in wind. This result is known under the name of *Betz's Law*. Bearing this limit in mind, and assuming normal conditions of temperature and pressure (15 °C, 1013 mbar), the maximum energy per unit of surface in W/m^2 which can be obtained from a wind of speed V is $P = 0.37\, V^3$. For a rotor of diameter d covering a surface S, Betz's limit gives an energy in watts as follows: $0.37\, SV^3 = 0.37(\pi/4)d^2 V^3 = 0.29\, d^2 V^3$. It is thus proportional to the cube of the wind speed and to the square of the rotor diameter of the wind generator. Bearing in mind the efficiency of the rotors and other components, wind

---

[6] During the daytime, the land warms up more quickly than the sea. The warmed air rises and draws in the cold air from the sea. This process is reversed at night when the Earth cools more quickly than the sea. Because of this, the wind blows from the land to the sea in the morning, when the surface of the sea is warmer, then from the sea to the land in the afternoon when the surface of the land becomes warmer than that of the sea.

generators currently available have an electrical efficiency, at nominal speed, of between 30% and 50% of Betz's limit. This efficiency also depends on the turning speed of the rotor.

Wind speed varies with the height above ground. It increases with altitude, but the variation depends on the nature of the terrain. It is preferable to site generators on a fairly even terrain to minimise the variation of wind speed along the blades and reduce the mechanical strains on them. As the wind speed at a height of 50 m can be 25–35% greater than that at 10 m, the energy yield doubles. This explains why it is preferable to have wind generators of a good height.

When a wind of constant speed strikes a surface, it creates a force whose value depends on the angle of incidence. This force is in two parts: one perpendicular to the wind direction (*thrust*), the other in the same direction (*drag*). For low incidences (<15°), the thrust grows more rapidly than the drag. For a wind generator blade, the thrust creates a torque that turns it in the place perpendicular to the wind. As for a sailing boat, the wind speed on the blades is the speed of the *apparent* or *relative wind*, which results from the real wind and the moving speed of the blade. The output (ratio between the recovered energy and the energy provided by the wind) depends, for each part of the blade, on the incidence of the wind.

The average speed of the wind 10 m above the ground surface in a site without obstacles generally varies between 1 and 10 m/s. Thus, the energy can vary from 1 to 1000, a considerable dynamic.

In France, wind power depends on the site [33]: 500 kWh/m^2/year in Toulouse, 1200 kWh/m^2/year in the Paris Basin, 6800 kWh/m^2/year in a very windy site like Força Real, near Perpignan.

A good quality wind is regular. Any rapid changes in speed or direction are disadvantages for wind generation. Turbulence is particularly damaging for wind generators because it reduces output and causes fatigue of the mechanical parts.

While wind energy is free, its recovery is not. Wind is intermittent and capricious, as sailors know too well. As the energy obtained is proportional to the cube of the wind speed, if wind speed doubles, the energy is increased eightfold. A modern wind generator needs a wind of at least 5 m/s (18 km/h, or Force 3 on the Beaufort Scale). For a wind generator of 750 kW, maximum output is obtained with a wind of 15 m/s (54 km/h, Force 7; at this speed, walking into the wind is difficult and trees are swaying). The output of this generator with a wind speed of 5 m/s is only 28 kW, that is very low. For a reasonable output, wind speed should not be below 11 m/s, or around 40 km/h (Force 6).

A speed limiter comes into force on a wind generator from 15 m/s, and the generator is halted when the wind speed reaches 25 m/s (90 km/h) to prevent damage. Some are designed to lie flat during violent windstorms like cyclones or hurricanes. The two-blade generators manufactured by Vergnet can resist wind speeds up to 300 km/h when lying flat.

There are two types of wind generator: those with a horizontal axis in relation to wind direction and those with a vertical axis. The former are more common because their efficiency is higher. Vertical generators are simpler mechanically but have a lower efficiency.

Wind generators are not as simple as they appear because they require quite complicated operating mechanisms. The materials used must be highly resistant because the stresses imposed by varying wind speed can be considerable.

The major problem with wind is its unreliability. This means that it is difficult to use a wind generator as the sole energy source without having facilities for storing the energy. A wind generator operates for 1550–2000 h/year at its nominal power output. In some sites, this can be extended to 3000 or even 3500 h.

The cost per kWh is still high, of the order of 5–12 euro cents, so this energy is subsidised. In France, where installed wind power is low compared to other countries like Germany, this subsidy represents a charge to the public purse of €600 million in 2010 if 5000 MW of wind generators are installed. However, wind generators can be competitive in some regions, like Guadeloupe for example. As with autonomous PV systems, wind generators are more expensive where they are not connected to the grid because of the need for batteries.

Large numbers of wind generators are needed to produce substantial amounts of energy. To produce energy equivalent to 1 GW of electricity, some 5000 wind generators of 750 kW are needed. The size effect does not work with wind generators because the linear velocity of the rotor tip has to be limited to keep the noise to a reasonable level, so a large rotor must turn more slowly than a small one. The only advantage of having large generators is that the number and visual impact can be reduced. Wind farms can also be sited offshore, but then there are problems of corrosion and limited access for maintenance.

Wind does not always live up to its promise and outputs are sometimes lower than expected. The German company EON reported that in 2004 the 14.3 GW installed in Germany had only produced 18.6 TWh, an output of only 14.8% and far from the 30% usually mentioned. [93]

Large wind turbines need to be connected to a grid, which is rather the opposite of what one expects – that a renewable energy source should be decentralised. This is because wind farms are built in places where the wind is strong and regular, and where the population density is generally low. The amount of electricity produced is far more than is needed to meet local needs and so has to be fed into the grid. The grid may have to be extended or strengthened to serve the wind farm.

The size of wind turbines is constantly growing and this will be particularly true of offshore installations. Wind generators of several MW already exist and there are plans for 10 MW turbines to be installed by 2010, with a rotor diameter of 160 m. That would seem to be close to the limit of the size that can be economically justified.

Wind generation must always be backed up by traditional production methods, since electricity must always be available even when there is no wind (which may be two-thirds of the time). A gas-fired power station is the most appropriate method of filling this gap. Major wind farms must always be backed up by gas-fired generation to supply energy when there is no wind. It might be thought that a very large wind farm covering a huge area would always have enough wind somewhere, but unfortunately that is not always the case, as has been shown in a number of heat waves or cold periods arising from a persistent anticyclone.

Wind generators are amply justified in a country like Denmark, for example, where a major part of the country's electricity is produced from coal. Wind turbines functioning around one-third of the time enable carbon dioxide emissions to be reduced by a third. However, in a country like France where electricity is produced mostly by nuclear energy, having to install gas-fired stations to make up for wind deficiencies (two-thirds of the time) would have the effect of increasing carbon dioxide emissions.

3.4 Biomass

Biomass is the energy stored in living matter such as vegetation. Using biomass to produce energy is an indirect way of using solar energy. Wood has been used since earliest times, perhaps 500,000 years, when humans discovered and used fire for warmth and light. Biomass is traded mainly outside traditional commercial channels although it makes up a considerable proportion of world energy consumption.[7] It is particularly used in many developing countries with a low population density. For 3 billion people – half of humankind – so-called traditional biomass (in the form of wood, plant waste, charcoal and dried dung) is the main if not the only source of energy [34]. It is often burned in a polluting open hearth with a very low efficiency ($\approx$10%). The number of users of biomass has increased considerably in the last 200 years from 1 billion in 1800 – total world population at the time – to 3 billion today. Biomass was extensively used in Europe in the past, and in France the massive use of firewood led to deforestation in the seventeenth and eighteenth centuries.

Plants accumulate energy during their growth in carbon–hydrogen chemical compounds through the process of photosynthesis that can be described in the following simplified equation:

$$n(CO_2 + H_2O) \xrightarrow{h\nu} (CH_2O)_n + nO_2 \tag{3.1}$$

Carbon dioxide and water are transformed under the action of light into carbohydrates and oxygen. Plants capture the intermittent solar energy and

---

[7] In France, consumption of commercial firewood for heating amounts to 2.5 million m^3 per year and 19.5 million m^3 for firewood sold outside commercial channels.

store it. The efficiency is low [35]. Only one-third of solar energy (the visible spectrum) is used in bioconversion and some of that passes through the foliage to the ground. More than eight photons are needed to reduce a mole of carbon dioxide in the photochemical process of higher plants and algae. The efficiency at cell level is therefore less than 30%. At the cultivation level, the maximum theoretical efficiency is less than 6%. Other natural factors reduce efficiency further, which in the case of sugar cane, for example, is only around 2%. In a temperate country like France, efficiency ranges between 0.4% and 1% for cultivated plants and is below 0.2% for forests. The solar flux arriving in France corresponds to around 1100 TOE/ha/year. With a conversion efficiency of 0.5%, this corresponds to an energy production of 5.5 TOE/ha/year, or 15 tonnes of dry matter per hectare per year.

Biomass is used for food, industry (furniture, paper pulp, etc.) and energy. Outside these applications it is also important for the protection of biological equilibrium and the fight against erosion, etc. Its main use is of course in food production. To place it in a proper perspective, energy fixed by photosynthesis corresponds to around 10 times the world energy consumption. Around 60% of it is in the form of terrestrial biomass and 40% marine biomass [35]. The quantity of energy represented by food is only equivalent to a twentieth of the world's total energy consumption.

As is shown by equation (3.1), biomass growth absorbs carbon dioxide, which is beneficial in relation to the growth of the greenhouse effect. But this is reversed when biomass is burnt. This would appear to even the balance, but the contribution of the greenhouse gases emitted in the manufacture of fertilisers must also be included, along with those created by the associated operations of transport, handling and processing. Still, the final balance is a positive one compared to the burning of fossil fuels.

The energy content of biomass is developed, for wet products by slow biochemical transformations and for dry products by thermochemical transformations that occur rapidly at high temperatures.

Half of all the wood humans use is for energy purposes. The area covered by forest in France is 14.6 million hectares – a quarter of the national territory; this represents 14% of the forest cover in Europe[8], and 0.3% of world forest cover. Between 3 and 15 tonnes of dry biomass can be produced per hectare [15]. Wood has a low water content compared to other plant matter – 40–60% for green wood, 20–25% for air-dried wood. Its calorific value is 14–18 MJ/kg when dry and 10–13 MJ/kg when air-dried (25% humidity). Taking the efficiency of wood-burning furnaces into account, 1 tonne of air-dried wood is equivalent to 0.25 TOE. In fact, wood does not actually burn. When it is heated, first it absorbs heat to eliminate any humidity, then at about 200 °C gives off gases – the phenomenon of *pyrolysis*. In the presence of air, these gases burn and give off heat and the temperature rises to around 800 °C. The wood is transformed into charcoal with a calorific value of 33 MJ/kg.

---

[8] Europe is defined here as the 15 countries in the European Union prior to 2004.

Biofuels can be produced from vegetable matter. At present they are more expensive than petroleum fuels, but can make an interesting contribution to the economy. They may be blended with traditional fuels in proportions of 5–10% or even more. It is also possible to produce ethanol or an ether derivative, ETBE (ethyl *tert*-butyl ether). They are either used as such (ethanol has been used as a fuel in Brazil since the 1970s) or blended with petrol. Diesel can also be replaced by methyl ester produced from rapeseed oil. The biofuels in use today are described as first-generation biofuels.

Biofuel productivity depends on the agricultural crop used. A hectare of wheat produces 2500 l of ethanol, whereas a hectare of beet produces 6500 l. A hectare of rapeseed produced 1300 l of ester. The energy balance, that is the ratio between the energy that can be recovered and that necessary for the manufacture of the biofuel, depends on the nature of the fuel and the type of crop. It is around 1.5 for bioethanol and 2 for rapeseed methyl ester, but the value depends on the crops used and the way the material is processed. Pessimists say that the gain is not worth the effort. Optimists see an amplification of energy choice in first-generation biofuels, and a home-grown fuel that supports local agriculture and reduces imports. A drawback is that the current enthusiasm for biofuels may have the effect of raising the price of other foodstuffs because of the competition for land use.

As far as ethanol is concerned, Brazil currently produces it at the lowest cost (around \$40/barrel oil equivalent in 2005) from sugar cane, a plant with a high energy output. Additionally, bagasse (sugar cane waste) can be burnt to produce part of the energy needed for the process. Biofuels have a positive effect on pollution, emitting less sulphur dioxide, fewer particles in the case of rapeseed ester and less carbon dioxide. On the other hand, there is increased emission of oxides of nitrogen because these fuels contain more oxygen than petrol, and the production of aldehydes. In 1999, 250,000 of ester and 200,000 tonnes of ETBE were produced from 350,000 hectares under cultivation. In 2004, France produced 504,000 tonnes of liquid biofuels, and this production level is still increasing. That same year, Brazil produced 12.3 Mt of biofuels and the world's total production was 30 Mt.

A European Directive of 8 May 2003 requires that the share of biofuels should be 2% in 2005 and 5.75% in 2010, shares calculated on the basis of its energy value. In 2005, the French Government undertook to reach the target of 5.75% in 2008, up from 1.2% in 2004. It set a target of 7% in 2010 and 10% in 2015. The yield of ester from rapeseed and sunflower oil is around 1 TOE/ha. The yields of ethanol are better (1.5–2 TOE/ha) for cereals (wheat, maize) and are as high as 3.5–4 TOE/ha for beet. But this crop needs a lot of water and other inputs and can only be grown in rich soils, and for best yields, the crop must be rotated.

Second-generation biofuels use wood-cellulose biomass (wood, straw, forest waste, etc.) that has the advantage of being grown without fertiliser. Plants that grow rapidly, such as coppiced trees and shrubs, provide the highest yields. The aim is to use the whole plant. After gasification of the biomass, fuels can be synthesised by the Fischer–Tropsch method. This process, developed in Germany during the Second World War and used more recently in South Africa, produces very pure (in particular sulphur-free) petrol. To increase the yield per hectare, the idea is to use an external source of energy that does not produce greenhouse gas emissions and to introduce excess hydrogen during the synthesis of the fuel in order to make use of all the carbon atoms contained in the biomass. Under research conditions, around 1 litre of fuel can be produced currently for every 6 kg of wood, but the aim is to be able to produce 1 litre of fuel oil for every 2 kg of wood within the next 5 years. The yield is much better than that of first-generation biofuels. One hectare of rapeseed produces around 1 tonne of first-generation biofuel, whereas the same area of wood-cellulose biomass can yield double or even triple the quantity of second-generation fuel.

The exploitation of biomass is worth pursing in France. It should be done by small units collecting biomass within a distance of 100 km. If the biomass is to make the best contribution against greenhouse gas proliferation, it should be replanted after being exploited – otherwise the biomass stock is diminished without being replaced, which is what happens with fossil fuels.

Translated into TOE, world food supplies only represent 500–600 MTOE, or some 6% of primary energy consumption. While in humankind's early days food was the largest energy input, today, at least in the developed countries, it represents a small part of our energy consumption.

We should also not overlook marine biomass whose yield per hectare can be up to ten times higher than that of terrestrial biomass.

3.5 Geothermal energy

The temperature rises as one penetrates deeper into the Earth's crust. This phenomenon is due to the geothermal energy flux. Although the Earth's core is very hot (around 4200 °C) and is slowly cooling, this is not the reason why the crust produces heat. The heat is essentially caused by natural radioactivity, mainly from uranium-235, uranium-238, thorium and the potassium isotope ^{40}K. Although the heat generated is low (18 J/g/year for U-235, 0.8 J/g/year for thorium or ^{40}K [37]), its accumulation in the Earth's crust is sufficient to generate heat that can be collected and stored in natural reservoirs like underground aquifers. In fact, 99% of the planet is hotter than 200 °C that explains why geothermal energy, although not strictly speaking renewable, is considered as such because it is virtually inexhaustible. Of course, this does not mean that a geothermal well will never become exhausted, but when it does, after several decades, another can always be drilled elsewhere.

The geothermal gradient, the rate at which the Earth's temperature increases with depth, is on average 3.3 °C per 100 m in the so-called normal zones. The flux of geothermal heat is around 0.6 W/m^2, or 3500 times less than the solar flux received on the Earth's surface [30]. In the Paris Basin, it varies between 3 °C and 5 °C, but there are areas where it is much higher. In Alsace, for instance, it is over 10 °C/100 m. Geothermal energy sometimes erupts from the ground in the form of geysers, as in Iceland. In regions where the tectonic plates are in collision, magma rises closer to the surface and high temperatures can be observed, reaching several hundred degrees at a depth of 1000 m. Geothermal systems may appear in regions where hot magma has risen near the surface (at a depth of 1000–1500 m). It may then directly heat an underwater aquifer or intermediary rocks. In certain conditions, hot water and steam from the geothermal system can reach the surface, hence geysers or hot springs are formed.

Geothermal energy was already exploited by the Romans who used the water from hot springs for heating, bathing and therapeutic purposes, for which hot springs are still widely used. The electricity from geothermal energy was first generated in Italy in 1903.

Geothermal energy sources are classified according to temperature and whether they can produce electricity or not [30].

'Low-energy' geothermal corresponds to temperatures between 30 °C and 100 °C. The heat sources are found from a few hundred metres to 2500 m depth. They are present in many parts of the world, particularly in sedimentary basins, such as the Paris Basin. If the water is not too corrosive, it can be used for heating, either directly or via a heat exchanger when the temperature or the pressure are too high. If the water contains high concentrations of mineral salts, it cannot be used directly in a heating system or discharged after use. It is pumped back underground, which has the advantage of maintaining the pressure of the source, but may lower the temperature. To avoid this happening, the used water is not reinjected at the pumping point, as was done at the plant at Melun near Paris. In France, low-energy geothermal is used for heating some 200,000 homes in the Paris and Aquitaine Basins corresponding to 1470 GWh/year. The investment costs of geothermal are around €400–600 per installed kW but the operating costs are very low: 0.05–0.1 euro cent per thermal kWh. When the temperature is low, around 30 °C, a heat pump may be used to extract the energy, which obviously needs to be powered by electricity.

'High- and medium-energy' geothermal can be used to generate electricity.

'Medium-energy' geothermal corresponds to hot water sources under pressure of temperatures between 90 °C and 180 °C. They may be at any depths from a few hundred metres to several kilometres. They can be used to run power stations ranging from a few kW to several MW, for an investment of between €1000 and €4000 per installed kW, and with a life of 30–50 years.

'High-energy' geothermal uses deposits of steam situated at a depth of 1500–3000 m at a temperature of 250–350 °C. When the steam is dry, which is unusual, it can be fed directly to a turbine to generate electricity. When it is wet, the steam must be separated from the liquid water at the top of the well. According to how the electricity is produced (from dry steam or after separation), the cost of the electrical kWh varies between 4.5 and 7 euro cents. The United States is the leading producer of geothermal electricity with 2700 MW installed. Geothermal generation in France, including the overseas territories, was equivalent to 117 GTOE in 2000 [11], including 4.2 MW installed in Guadeloupe. In 2004, the world production of electricity from geothermal was 55.9 TWh, and the amount of heat generated was 14,900 TJ [91].

Geothermal energy has a low impact on the environment. It emits little carbon dioxide, but there are emissions from gases dissolved in the water, such as methane, hydrogen sulphide and some carbon dioxide. The waste water from geothermal exploitation cannot be discharged at the surface because it contains salts and heavy metals. It is normally reinjected underground.

The energy contained in the heat of the soil not far below the surface, and indeed any other source of low-level heat, can be recovered by heat pumps. This process is under-utilised used in France but has interesting future prospects. Using a heat pump, it is possible to generate around 4 kWh of heat using only 1 kWh of electricity, the remaining energy being recovered from the low-temperature heat source as it cools. This method has applications far beyond geothermal and can recover energy from many different sources – seas, rivers, heating networks, etc.; the heat pump may be seen as an energy amplifier.

3.6 Energy from waste

The energy that can be recovered from waste may be considered renewable, because as long as there are humans there will be waste to be processed. The first aim of waste processing is to reduce the volume of the waste and remove the noxious elements of waste, especially health hazards. When heat is given off or when energy is produced in the course of waste treatment, it is obviously in the public interest to collect it and exploit it. Organic waste is obviously bio-mass. Two main processes are used to recover energy from waste: *fermentation* and *incineration*.

Fermentation uses bacteria that decompose the organic matter either in the presence of air (aerobic process) or in the absence of air (anaerobic process). The first process is involved when household rubbish is discharged or when waste water is treated. This type of fermentation is exothermic – in other words, it releases heat.

Anaerobic fermentation can be more easily controlled because it takes place in a digester. It is quicker than aerobic fermentation. For example, while the aerobic composting of garden refuse takes several months, anaerobic fermentation can do the same job in 2–3 weeks [30].

Fermentation produces biogas, a mixture of carbon dioxide (20–50%) and methane in proportions that depend on the nature of the waste, and a residue that can be used for soil improvement in agriculture. Anaerobic fermentation produces little heat, and therefore some energy has to be used to speed the reaction that takes place at 35 °C or even 55 °C. Part of the biogas produced (25–80%) is used for this purpose. Apart from carbon dioxide and methane, biogas also contains hydrogen sulphide (H_2S) and other products. The water vapour present leads to the formation of corrosive acids, hence stainless steel or plastic must be used for the vessels.

The main wastes treated by anaerobic fermentation are animal waste from pigs, cattle and chickens, industrial waste (fodder industries, paper mills), household waste or sewage sludge. Animal manure is still an important source of energy in countries like China and India. In Europe, animal waste is eight times the size of household waste.

Depending on its end use, biogas has to be purified to remove products like hydrogen sulphide. Mostly, it is used in installations close by to supply heat, particularly to help run the digester. Biogas can also be used to generate electricity alone or with heat (co-generation). The larger the installation, the lower the price of the electricity – typically it will be around $4-8$ euro cents. One tonne of household waste can be used to produce $100-200$ m^3 of biogas over 3 years [30]. Biogas can also be used, like pure methane, as a fuel for motor vehicles.[9]

Incineration is another method of reducing waste. It can be used for household waste, agricultural waste (straw, paper mill residues, etc.) and for some industrial waste (used solvents, tars, etc.).

Each European produces on average 1 kg of household waste per day – a US citizen twice as much. Around 70% of this waste is combustible. The rest – glass, metals – is not. Incineration can reduce the weight of waste by 70% and the volume by 90%. With 5–7 tonnes of household waste, energy equivalent to 1 tonne of oil can be produced. The amount of energy depends on whether heat or electricity is being generated. Thus, 1 tonne of waste can produce 1500 thermal kW but only 300–400 electrical kWh. Co-generation is also possible. Although it is worthwhile to produce energy from household waste, the quantity that could be produced in France only amounts to about 1% of total energy consumption.

Agricultural waste apart from wood includes straw, waste from agro-food industries like sugar factories and oil mills and the black liquor from paper mills. One hectare of cereal crops produces around 4 tonnes of straw, the calorific value of which is 17 MJ/kg when dry, less when wet. Straw thus has the same calorific value as wood but occupies, for the equivalent energy content, a space 4–8 times larger. Electricity can be generated from it at a price hardly more than that of conventional methods. Agro-food wastes like bagasse – sugar cane residue – have a calorific value similar to wood when dry. It is now possible to produce 1 kWh with 2 kg of bagasse at a cost of around 8 euro cents.

Some organic industrial wastes, like bone meal, can be treated in high-temperature incinerators (over 1000 °C) or as fuels to replace oil in cement works, for example.

3.7   Conclusion

There is now increased interest in renewable energies because of technological progress. They are valuable in certain niche markets such as isolated settlements, off-grid locations and overseas territories. They are indispensable in developing

---

[9] Although highly compressed when used as a fuel for vehicles, methane occupies five times as much space as a liquid fuel, which limits the vehicle's range (120 km for a tank of 50 l).

countries with low population density. They are not yet fully exploited for heat production and more should be done to develop this potential.

Figure 3.3 shows that the electricity produced in France in 1998 from renewable energy was mainly derived from hydropower. Wood and waste also produced more electricity than solar or wind power at that time.

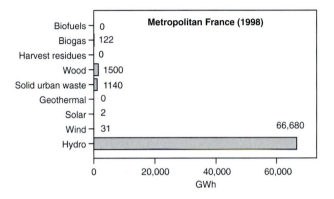

Figure 3.3 Production of electricity (in GWh) from renewable sources in metropolitan France in 1998 [Source: DGEMP, Energy Observatory]

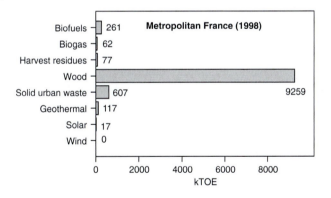

Figure 3.4 Production of heat (in kTOE) from renewable sources in metropolitan France in 1998

Figure 3.4 shows the predominance of biomass over other renewables for the production of heat in France in 1998. Adding the quantity of energy to the heat produced, there was a total of 25.8 MTOE (26.1 MTOE including the overseas territories) produced from all renewables. That represents a tenth of 1% of the total energy consumed in France, which is nearly double the European average, and these conclusions are still valid today.

Biomass is far from negligible. Even today it represents 4.5% of primary energy consumption in France, without counting all that is used outside the traditional commercial markets. Apart from hydropower, the other renewables only represent 0.05% of French energy consumption. Biomass is therefore an asset to the French economy and other uses for it are likely to be found.

In France, the flattening of electricity prices means that the price of the kWh is the same everywhere and all the time, which works against renewable energies. Some would be more viable in certain situations if this price mechanism did not apply.

When a renewable energy is economically competitive and convenient, there is no resistance to its use. This is the case with hydropower and some forms of biomass. Unfortunately, most are still far from being economically competitive, which justifies more research into lowering their cost and making them easier to use. Without a technological breakthrough, the market penetration of renewables will be slow. Even if their use in absolute terms has increased, in relative terms it is not because of the growth in total energy demand. J.R. Bauquis [38] has calculated that at the global level, the number of renewable GWh of electricity produced (excluding hydropower) will grow from 100 to 1250 GWh between 1995 and 2050 (from 2500 to 4250 GWh including hydropower). This corresponds to the contribution of renewables to electricity generation rising from 0.8% to 3% if hydropower is excluded, but falling from 19% to 10% if hydropower is included. For total energy consumption, the share of renewables will grow from 7.5% to 8% between 2000 and 2050 [38].

It is important to develop renewable sources of energy that are best suited to a particular country in terms of economy and employment. There is no universal solution, and the renewables to be most developed in France are not necessarily the same as those of our neighbours. For instance, France has a lower population density and a higher agricultural coverage than some of our neighbours, which therefore makes the development of biomass use particularly appropriate.

One important point is that more needs to be done to inform and educate the public about the benefits of renewable energy, and more young people need to be trained in trades associated with their application. At present, in France, there is a shortage of qualified professionals trained in the installation and maintenance of most renewable energy equipment, so there is little chance of seeing a rapid increase in their use.

It is essential to develop the renewable energies because their use is important to the sustainable development of our planet. They are particularly indispensable for many people who live in a state of energy poverty (some two billion people have no access to electricity). These disadvantaged people should be given priority access to renewable energy sources and financial mechanisms should be put in place to enable this to happen.

In the absence of significant technological progress, which seems unlikely, renewable energies will only make a small contribution to global energy consumption, and will not easily supplant other forms of energy. A significant contribution will no doubt have to wait until the second half of this century. The energy sector changes slowly, and what is being planned in today's laboratories may only be operational several decades hence. Time and heavy investment are also needed for a new product to penetrate the market.

Chapter 4

Nuclear energy

Knowledge of atomic physics is comparatively recent. Natural radioactivity was discovered in 1896 by Becquerel. In 1934, Frédéric and Irène Joliot-Curie created the first artificial radioactive element. Nuclear fission was discovered by Otto Hahn and Fritz Strassmann in 1938. In 1939, Frédéric Joliot, Hans Alban, Lev Kowarski and Francis Perrin demonstrated that it is possible to initiate a fission chain reaction to produce energy, and the principle of the nuclear reactor was born.

The combustion of oil, coal or gas is a chemical reaction that produces heat by reorganising the electronic shells of atoms. The energy liberated by a simple reaction, that is only affecting the atoms necessary to achieve it, is measured in electron volts (eV, 1 eV = 1.6×10^{19} J).[1] An atom is made up of electrons and a nucleus composed of nucleons. Nuclear reactions correspond to a reorganisation of nucleons in the interacting nuclei. Some nuclear reactions, such as the fission of certain nuclei, produce energy. The energy liberated, for an equivalent mass, is more than a million times greater than a chemical reaction and is measured in MeV – millions of electron volts.[2] This enormous concentration of energy has the important consequence that nuclear waste, by volume, is a million times more concentrated than those arising from a chemical reaction. Some of these waste products are dangerous for humans, but their small volume means that solutions can be found for their temporary storage and disposal.

Nuclear reactions can produce considerable quantities of energy at a cost highly competitive with other sources. Anything to do with nuclear energy and radioactivity can generate irrational fears that no doubt partly originate from the use of the atom bomb in 1945. As radioactivity is invisible, there is great suspicion of nuclear phenomena, although these have existed from the beginning of the universe, and enable us to exist on earth through the presence of the Sun, which is powered by nuclear reactions.

---

[1] The combustion of a carbon atom ($C + O_2 \rightarrow CO_2$) liberates around 4 eV.
[2] The fission of a nucleus of uranium liberates around 200 MeV.

4.1   Atoms and nuclei

All natural objects, whether living or inert, are composed of atoms. Atoms themselves are made up of three types of particles: protons and neutrons, which form the nucleus, and electrons. Neutrons and protons belong to a family of particles called nucleons, whose mass is rather more than 1800 times that of the electron. An atom is described by the number of its protons, Z, which is called the atomic number, and its mass number A, which is the total number of nucleons (protons and neutrons) contained in the nucleus. The number of neutrons therefore equals A–Z. The neutron is electrically neutral, whereas the proton carries a single positive charge $+e$. The protons and the neutrons of the nucleus are bound by nuclear forces that are very intense but only act over a short distance. The charge of the nucleus of an atom of atomic number Z and mass number A is $+Ze$. To balance this positive charge and obtain an atom that is electrically neutral, the atom possesses Z electrons, as electrons carry a negative electrical charge of $-e$. The nucleus and electrons are bound by electromagnetic force that is 100 times weaker than the nuclear force, but whose range is infinite.

The mass of an atom is concentrated in the nucleus, whose dimensions are a few femtometres (10^{-15} m). The dimensions of an atom are a few angstroms (Å) (1 Å $= 10^{-10}$ m), or 10^5 times larger than the nucleus. Thus, an atom is essentially made up of empty space. On the other hand, the density of matter in the nucleus is enormous: 2×10^{17} kg/m^3 or 200 billion kg/cm^3.

The chemical properties of an atom are determined by the number of its electrons, equal to its atomic number Z. Atoms having the same number of protons but different numbers of neutrons are called isotopes. These have the same chemical properties but different nuclear properties. For example, there are three isotopes of hydrogen:

1. 'Ordinary' hydrogen, the nucleus of which contains a single proton, which makes up 99.9% of naturally occurring hydrogen.
2. Deuterium, the nucleus of which is made up of one proton and one neutron. It represents 0.015% of naturally occurring hydrogen.
3. Tritium, the nucleus of which is made up of one proton and two neutrons. It does not exist in the natural state because it is unstable.

It is usual to describe an atom X made up of Z electrons and a nucleus containing A neutrons (Z protons and A–Z neutrons) as $_Z^A X$.

4.2   Radioactivity

Many natural and artificial nuclei are unstable. They have an excess of energy and can evolve (or disintegrate) to become more stable, sometimes passing through intermediate states that are themselves unstable. During this process, they generate another nucleus (which may be the same but in a lower energy state) by emitting a particle, for example an electron, a helium nucleus, sometimes a photon (gamma) or by breaking into two parts, as in fission. This is the process called radioactivity, and the nucleus is said to be radioactive. Radioactivity can be natural or artificial, according to whether human intervention is behind it. More than 1500 artificial radioactive isotopes have been created, but the great majority of the radioactivity present on earth is of natural origin.

The energy liberated during radioactive decomposition is divided between the various components found in the final state. The three principle products of disintegration are α, β and γ radiation. In alpha radioactivity, the initial nucleus of atomic number Z and mass number A ($^A_Z X$) emits a nucleus of helium $^4_2$He, composed of two protons and two neutrons:

$$^A_Z X \rightarrow ^{A-4}_{Z-2} Y + ^4_2 \text{He} \tag{4.1}$$

This radioactivity occurs particularly with heavy nuclei. The major part of the energy liberated is in the form of kinetic energy, that is, in the movement of the helium nucleus, also called the alpha particle, and the remaining daughter nucleus, $^{A-4}_{Z-2} Y$. As the alpha particle is much lighter than the nucleus, it carries the major part of the available kinetic energy.

There are two types of β radioactivity: $-\beta^-$ and β^+. In β^- radioactivity, an electron e^- (also called β^- ray) is emitted with an antineutrino ν^-, a particle of negligible mass and without a charge, which carries, in kinetic form, part of the energy liberated during the reaction.

$$^A_Z X \rightarrow ^A_{Z+1} Y + e^- + \nu \tag{4.2}$$

The resulting nucleus advances one place in the periodic table compared to the initial nucleus. In β^+ radioactivity, a positron (anti-particle of the electron) e^+ is emitted together with a neutrino,[3] which are formed during the reaction:

$$^A_Z X \rightarrow ^A_{Z-1} Y + e^+ + \nu \tag{4.3}$$

The resulting nucleus goes back one place in the periodic table.

Gamma radioactivity corresponds to the emission of electromagnetic radiation of very short wavelength (gamma photons are very energetic light particles) by a nucleus that is in an excited nuclear state. It is the equivalent, for a nucleus, of the emission of radiation by an atom when an electron passes

---

[3] Neutrinos and antineutrinos are elementary particles without charge and with negligible mass that only interact weakly, which means they are practically undetectable. To stop the considerable flux of neutrinos arriving from the Sun ($\approx 65 \times 10^9$/cm^2/s) would need a wall of lead with the thickness of a light-year (10^{16} m). Postulated in 1930 by Fermi to explain the phenomenon of beta decay, their existence was only proved in 1956.

from a high to a low atomic level. The mass and atomic number of the nucleus remain unchanged:

$$^A_Z X \rightarrow ^A_Z X + \gamma \tag{4.4}$$

For the energy of alpha particles, electrons (e^-, e^+) and gamma rays detected during radioactive decomposition, it is observed that in general:

- alpha radiation is stopped by a few centimetres of air, a sheet of paper or skin;
- beta radiation is stopped by a few metres of air, a sheet of aluminium foil or one centimetre of human tissue;
- gamma radiation can travel through several hundred metres of air and beyond. It goes through a human body and requires around 10 centimetres of lead to stop it.

All radioactive nuclei disappear sooner or later. It is a random process and each nucleus has a certain probability of decomposing over a period of time. It is usual to describe it by its *period* $T_{1/2}$, which is the time that a radioactive source made up by nuclei of this element takes to lose half of its activity. In other words, a number N of radioactive nuclei has a half-life $T_{1/2}$ if at the end of time $t = T_{1/2}$ there only remains half of the radioactive nuclei ($N/2$). At the end of two periods ($t = 2T_{1/2}$), there remain $N/4$ radioactive nuclei. This period of half-life is a characteristic of radioactive nuclei. The half-life can be less than one second or more than one billion years.

The radioactive nucleus has no age; it always remains young. Its probability of disappearing always remains the same as long as it exists. A phenomenon that does not depend on the past is called a *Markov process*. It is the opposite for a human being, who ages as time passes.

The unit of measurement of radioactivity is the *becquerel* (Bq), which describes the activity of a quantity of radioactive material in which one nucleus decays per second. It is a very small unit.[4] It replaced the *curie* that was a very large unit (3.7×10^{10} Bq or 37 billion Bq).

Radioactivity was not invented by human. It has been present since the existence of nuclei, that is shortly after the creation of the universe 14 billion years ago. It has lessened in the course of time, and is becoming weaker all the time. The intensity on the Earth is less now than when the Earth was created 4.5 billion years ago. The average radioactivity of the earth's crust has fallen from 5840 Bq/kg at that time to 1330 Bq/kg now [39]. The radioactivity of uranium has fallen by half, and that of ^{40}K has fallen from 4230 to 370 Bq/kg. We will look at the impact of radioactivity on the environment and living beings in Chapter 6.

---

[4] The natural radioactivity of a human body is about 8000 Bq.

4.3 Mass and energy

At the birth of the universe, energy was transformed into matter. We know now that matter can also be transformed into light. This is what happens when an electron e^- meets its anti-particle, the positron e^+. Both are destroyed, and two gamma photons are created. Einstein demonstrated in 1905 that energy e of a body of mass m at rest was equal to:

$$E = mc^2 \qquad (4.5)$$

where c is the speed of light (around 300,000 km/s).

A change in energy ΔE may be accompanied by a change in mass ΔM and vice versa, which is why we can say that there is equivalence between mass and energy. This is expressed in the relationship:

$$E = m \cdot c^2 \qquad (4.6)$$

Thus, the mass of an atom made up of electrons and a nucleus is smaller than that of the nucleus and its electrons taken as free particles. The difference is the binding energy that acts as a glue between the electrons and the nucleus and within the nucleus itself. The binding energy corresponds to a conversion into energy of an almost infinitesimal mass. The binding energy of an atom of hydrogen is 13.6 eV, or a change in mass of 2.4×10^{-22} g, since:

$$1\,\text{kg} = 5609 \times 10^{29}\,\text{MeV} = 2.5 \times 10^{10}\,\text{kWh} = 25\,\text{TWh} \qquad (4.7)$$

The fission of a nucleus of U-235 releases approximately 200 MeV, corresponding to a reduction is mass of 3.6×10^{-25} g, between the initial state and the final state. This value is low, but becomes more significant when the energy production of a whole country is considered. France produced 450 TWh of electricity from nuclear power stations in 2006 [11]. This amount of electricity represents the transformation of a mass of 18 kg into energy. In fact, the mass transformed is larger, because power stations have an efficiency, according to Carnot's principle, of 33%. Consequently, two-thirds of the energy produced by a reactor is lost in heat, and more than 50 kg of mass is used in the generation of electricity in France every year.

The amount of binding energy per nucleon of natural nuclei is shown schematically in Figure 4.1. The most stable nuclei are in the area of iron, whose binding energy is close to 8.7 MeV/nucleon. The curve is relatively flat around the maximum. The nucleus of helium is unusual among light nuclei because it is particularly stable (7.1 MeV/nucleon) compared to others. The figure shows that heavy nuclei are less bound than those of medium mass (7.7/MeV/nucleon for uranium). Consequently, if a heavy nucleus is split into two nuclei of medium mass, energy is produced. This is what happens during the fission of uranium, which liberates rather less than 1 MeV/nucleon.[5]

---

[5] The energy liberated is around 200 MeV for around 236 nucleons.

We can also see that the fusion of two light nuclei in helium liberates a lot of energy, generally several MeV/nucleon. Fission and fusion are therefore two different routes for producing energy. Fusion is still in the research phase and will be dealt with in Chapter 7.

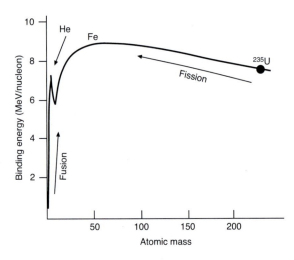

Figure 4.1   Schematic representation of binding energy (in MeV/nucleon) of stable nuclei as a function of mass

4.4   Fission

Some heavy nuclei split spontaneously into two fragments of medium mass, accompanied by neutrons, while others do so after having absorbed a neutron. This is the phenomenon of fission. Uranium-235, which is an isotope of uranium made up of 92 protons and 143 neutrons, has a large probability of capturing a neutron whose speed corresponds to the thermal agitation of the medium. Although called *slow* or 'thermal', these neutrons proliferate at 20 °C at a speed of 2.2 km/s, which corresponds to an average kinetic energy of 1/40 eV. The nucleus of U-236 that is created is excited ($^{236}U^*$) and splits into two pieces of medium mass called fission fragments, and, depending on the circumstances, two or three neutrons are emitted. The energy liberated is around 2000 MeV (30 pJ):

$$^{235}U + n \rightarrow\ ^{236}U^* \rightarrow 2 \text{ fission fragments} + 2 \text{ or } 3 \text{ neutrons} \qquad (4.8)$$

Some fission reactions release two neutrons, others three, and the average is 2.5. Each neutron carries an average kinetic energy of 2 MeV (or 5 MeV altogether). The neutrons emitted are *fast*: their speed is around 20,000 km/s.

The sum of the kinetic energy of the fission fragments is of the order of 165 MeV. The gamma rays and the antineutrinos each carry a dozen MeV and the electrons (beta) 8 MeV.

Not all the neutrons are emitted at the same time. Many come from the de-excitation of the fragments of fission formed (primary fragments). They are emitted very rapidly and so they are called *prompt neutrons*. However, by radioactive decay some primary fragments can evolve into other excited isotopes that may in turn be de-excited by the emission of neutrons. This emission can take place much later than that of the prompt neutrons (this is linked to the period of radioactive decay of the nucleus that has led to these secondary fragments) hence the name *delayed neutrons*. The proportion of these is small: 0.65 per 1000 for U-235 and 0.1 per 1000 for Pu-239. But they play a key role in the control of nuclear reactors; if they did not exist, the reactors could not function.

4.5 Chain reactions

In an environment containing fissile nuclei in which a fission reaction has been initiated, some of the neutrons liberated may initiate a further fission. The others may escape from the environment, be captured by fissile nuclei without causing fission or by other nuclei present in the environment. This leads to a definition of the neutron multiplication factor k as the ratio of the number of fissions of the generations $n + 1$ and n:

- If $k > 1$, the number of neutrons grows exponentially, and the environment is described as *supercritical*. As the time separating two generations of neutrons is very short, a divergent evolution quickly results. The system behaves like an amplifier; a small quantity of initial energy generates a lot of energy. This is what happens in an atom bomb of type A, where the environment is confined as much as possible in order to generate the greatest amount of energy before it is liberated.
- If $k < 1$, the environment is said to be *subcritical* and the fission reactions stop spontaneously.

In a nuclear reactor, the population of neutrons is controlled to extract energy in a continuous manner, and k is only very slightly superior to 1 at the start. Once the nominal power has been reached, the reactor is controlled to remain in the vicinity of $k = 1$.

The value of k depends on the geometry of the environment (if it is small, many neutrons could escape to the exterior), on its composition (some nuclei can strongly absorb neutrons and act like poison towards the reaction) and on its heterogeneity (the use of different materials in the system has a direct influence on the properties of the neutron cloud), etc.

The neutrons captured by a nucleus may lead to a radioactive isotope or in some cases to a fissile isotope. This happens for example with nuclei of

Uranium-238 and Thorium-232 that are transformed by the capture of a neutron followed by a double beta decay, resulting respectively in Plutonium-239 and Uranium-233, both fissile materials. A nucleus that develops by capturing a neutron into a fissile nucleus is termed *fertile*. Uranium-238 and Thorium-232 are fertile nuclei from which it is possible to create fissile nuclei and hence more fuel. This is the principle of a *breeder* reactor where, starting from a fissile–fertile mixture, the fissile part can be used to produce energy and transform part of the fertile nuclei into fissile nuclei through the neutron flux existing in the reactor. At the end of the operation, there are more fissile nuclei than at the start. The elements consumed are the fertile nuclei.

4.6 Nuclear reactors

A nuclear reactor is a system in which a fission chain reaction sustained by neutrons is maintained. Its structure differs according to the output required.[6] Here, we shall consider nuclear power stations that produce electricity. Their advantage is the ability to produce large quantities of heat at low cost. This heat is used to produce steam, which drives a turbine group similar to those that are used in traditional power stations burning fossil fuels.

A nuclear reactor functions according to simple principles where the safety of the plant is an essential consideration for builders and operators. The aim is to use the fission induced by neutrons from a fissile nucleus. As seen above, a chain reaction can be established and energy can be recovered. For this to happen, too many neutrons must not be lost between two generations, as some of them can escape from the environment or be captured without leading to fission.

Uranium-235 fissions with neutrons of all energies, but with slow neutrons with energy below 2 eV the probability of fission is very high. Uranium-238 only fissions with rapid neutrons with energy that is above 1 MeV and absorbs most of the neutrons of intermediate energy. There are two main families of reactors: reactors with slow or thermal neutrons and reactors with fast neutrons.

Slow neutron reactors or *thermal reactors* make use of the high probability of Uranium-235 capturing thermal neutrons. As the neutrons emitted by fission are mainly rapid, they have to be slowed down by a *moderator* material. The slowing takes place during elastic collisions of the neutrons with the nuclei of the moderator. A good moderator should absorb neutrons only weakly (transparency) and consist of light nuclei so that the slowing is as efficient as

---

[6] A nuclear reactor can be designed to produce:

- bundles of neutrons for scientific, technological or medical applications,
- heat for urban or industrial heating,
- mechanical energy for naval propulsion,
- electricity.

possible. Heavy water offers a good compromise between slowing and transparency (35 impacts on average to thermalise the neutrons), which allows natural uranium to be used as a fuel. Light water is more efficient in terms of slowing (19 impacts on average) but absorbs more neutrons, which necessitates the use of a fuel containing slightly enriched uranium.

Graphite can also be used. This is less efficient in slowing the neutrons, but only absorbs them weakly. On average, 115 impacts are needed to thermalise neutrons.

If natural or weakly enriched uranium is used, a reactor has a divided structure in which moderator and fuel are well separated to minimise the capture of neutrons by Uranium 238. If a fuel enriched with Uranium-235 is available, a more compact structure is possible, where the moderator is placed nearer the fuel.

Fast neutron reactors have no moderator. They use a high proportion of fissile nuclei because the probability of fission is much lower.

Once it has reached its nominal power, a nuclear reactor must work in a state where the number of neutrons inducing fission of the generation, $n + 1$, is equal to those of generation n. This is the critical state, and the multiplication factor is $k \approx 1$. If there were only prompt neutrons that are emitted in a very short time, the system would diverge rapidly. For the French reactors, this time is 2.5×10^{-5} s or 40,000 generations of neutrons per second [40]. Where $k = 1.0005$, the power would be multiplied by nearly 483 million in one second! There would not be time to control the reactivity of the reactor. Luckily, the existence of delayed neutrons, some of which are emitted several tens of seconds afterwards (11 s on average) makes it possible to control the chain reaction. This leads to a larger apparent period and a period of divergence of the order of 85 s [41]. Thus, a reactor must always be subcritical in prompt neutrons but may be slightly supercritical in delayed neutrons.

The reactivity of an environment is defined by:

$$\rho = \frac{k - 1}{k}$$

The system is stable if $k = 1$. The divergence from stability is measured in units of 10^{-5} called pcm (pcm = per cent mille). Thus, $k = 1.001$ corresponds to a reactivity of 100 pcm. For a thermal uranium reactor, it is essential that $\rho < 650$ pcm in order to have sufficient delayed neutrons to operate the reactor and remain subcritical in prompt neutrons. For a plutonium reactor, $\rho < 360$ pcm, which is more limiting.

Control of the reactor reactivity [42] is carried out with control rods containing elements that absorb neutrons (cadmium, boron, etc.), and with the help of various devices such as, for pressurised water reactors, water containing boric acid.

Several mechanisms enable the state of the nuclear reactor to be returned naturally to its position of stable functioning should it diverge slightly from this [42].

4.7   Natural reactors

The first nuclear reactors were not made by human, but by nature. This conclusion was reached in 1972, after a CEA laboratory found samples of natural uranium coming from the Oklo uranium deposits in Gabon whose content of Uranium-235 was below the 0.7% normally observed (as low as 0.4%).

The modification of the isotope composition originated in natural reactors that started spontaneously about two billion years ago when particular conditions were met. The radioactive half-life of Uranium-238 is 4.5 billion years and that of Uranium-235 is 700 million years. Uranium-235 therefore disappears more quickly than uranium 238. Two billion years ago, the isotopic concentration of Uranium-235 was higher, around 3.6%, which is similar to that used for the fuel of pressurised water reactors [41]. In that part of the world, the concentration of uranium ore was also high (more than 10%), the terrain is very porous and has high rainfall. Deep underground, pressure and temperature were high, and there were conditions similar to those in today's pressurised water reactors (PWRs).

Several natural reactors started spontaneously and functioned for a long time (probably for between 10,000 and 80,000 years), until the fuel was exhausted, leaving unprotected waste in the soil. Some of this waste migrated and some were retained, like plutonium that has now disappeared, as its half-life is only 24,000 years. The Oklo site is thus a mine of information because what is there is a store of unconfined nuclear waste, a situation that would be impossible today.

4.8   Nuclear power stations

The first nuclear reactor was created by Enrico Fermi and his team in 1942, beneath the benches of a Chicago stadium. Its power was very low -0.5 W. Electricity was produced for the first time in 1951 at the Idaho Falls site in the United States, but it was only in 1954 that the first nuclear reactor was connected to an electric grid at Obninsk near Moscow. In France, the first reactor using the UNGG process (Uranium Naturel, Graphite, Gaz), with a power of 40 MW, was commissioned at Marcoule in 1956. In 2006, 441 nuclear reactors, with a total electrical output of around 380 TW, have produced more than 2600 TWh of electricity, which represents in oil equivalent nearly 0.7 GTOE, or slightly more than 16% of total electrical production, and nearly the equivalent of that generated by hydropower.

Many combinations of fuels, moderators, heat transfer liquids and architectures may be used to design a nuclear reactor. In theory, more than 100,000 combinations are possible. About 1000 have been studied and of those around 100 have been seriously studied [43]. Several dozen were actually constructed, between 1950 and 1965. Gradually, a selection was made and certain methods have taken over from others for technological, economical or political reasons.

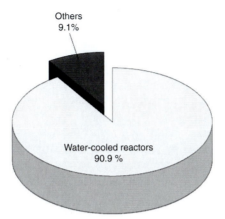

Figure 4.2   Water-cooled reactors (PWR, BWR and VVER) dominate the
market [Source: ELECNUC, CEA, 2006]

Light water reactors dominate the market (see Figure 4.2), making up more than 90% of all nuclear power stations. Those in France are PWRs. Water is both a very good moderator because hydrogen is the most efficient nucleus for slowing neutrons, and it also plays a role in heat transfer, removing the heat produced by fission. These properties make for very compact reactors. As hydrogen also captures neutrons (650,000 times more readily than deuterium), it is necessary to use uranium enriched at least 3% in Uranium-235, or plutonium. A prerequisite is therefore an ability to purchase or manufacture enriched uranium. There are two variants of water reactor [40]:

1. Pressurised water reactors (PWRs), where the water of a primary circuit in contact with the fuel rods is maintained under high pressure (155 bar) to avoid boiling. The heat recovered is transferred to a secondary water circuit that produced steam, which drives a turbine that produces electricity by means of an alternator. The water of the secondary circuit is cooled, condensed and recycled. The cold source may be water from a river, the sea or a refrigeration tower. The PWR technology developed in the Soviet Union is called VVER. At the end of 2005, there were 241 GWe of PWR installed globally (267 reactors) and of those 36 GWe are from VVER (53 reactors).
2. Boiling water reactors (BWRs), where water is boiled and used directly to drive turbines. The design is simpler than with PWRs because there is only one circuit. However, for a given power, the volume of the core is larger because the water is not under pressure. There volume power is 50 W/cm^3, whereas it is 100 W/cm^3 for a PWR [45]. There were 78 GWe of BWR installed (89 reactors) at the end of 2006 [44].

Heavy water reactors are a Canadian speciality. This process, called CANDU (Canada-Deuterium-Uranium), uses heavy water as a moderator and heat transfer liquid. As deuterium absorbs very few neutrons, it can use a fuel of natural uranium oxide that avoids dependence on the uranium enrichment industry.

The graphite gas process uses graphite as a moderator and CO_2 compressed to around 40 bar as coolant. These reactors are large because a lot of graphite is needed to slow the neutrons. The first French nuclear power stations were of this type, but this process was abandoned in France for the PWRs, which are more economically competitive.[7] The installed power worldwide is close to 11 GWe with 22 reactors [44].

The RBMK (Reaktor Bolshoi Moshchnosty Kanalny) process deserves particular attention because the Chernobyl reactor, which was the site of a serious accident in 1986, belongs to this category. These are reactors functioning with lightly enriched uranium oxide, cooled by boiling water and moderated by graphite. Unlike all other reactors in the world, these have the disadvantage of becoming unstable in certain conditions.[8] This type of reactor has no containment structure as exists for today's Western reactors. In 2005, there remained around a dozen reactors of this type in the world.

Good availability of nuclear power stations is an important source of economy. For instance, there was a fleet of 103 reactors in service in the United States in 2000, which produced 754 TWh. Although there were about 10 reactors fewer than in the 1990s, an improvement from an average availability of 70% in the 1990s to 89% in 2000 enabled an increase in electrical generation equivalent to the commissioning of 23 new reactors of 1000 MWe.[9]

In France, an increase in nuclear reactor life from the 30 years planned to 40 years would enable the cost of the kWh generated to be lower, because the amortisement of the investment was calculated over 30 years. For the 10 extra years, there would only be the operational costs.

4.9   Nuclear fuel

The only fissile nucleus existing in nature is U-235, which represents 0.7% of natural uranium. Uranium-238 that makes up 99.3% of natural uranium is fertile, that is it can be transformed into a fissile nucleus Pu-239 after the capture of a neutron and a double beta decay. The U/Pu couple forms what is called the uranium/plutonium cycle that is used in the nuclear industry. Thorium could also be exploited and its reserves are around four times greater than those of uranium, but the Thorium-232 present in nature is fertile but not fissile. After capture of a neutron, it leads to fissile U-233. The

---

[7] There is also a modern high-temperature version of gas reactor using helium as coolant.
[8] This instability, associated with an intentional short circuit during an experiment, led to the accident at Chernobyl which is discussed in Chapter 6.
[9] 1 MWe is 1 MW of electricity. To generate it, a higher power is generally needed because of losses (1 PWR of 3 GW generates 1 GWe).

thorium/uranium cycle could be used in the future when there is no longer any economically exploitable uranium.

In the Earth's crust, there are on average 3 g of uranium per tonne.[10] Uranium is exploited in regions where the concentration is much higher than the average. Mines exist in France, but these are virtually exhausted or not economically viable. The largest known deposits in the world are in Canada, Australia and Kazakhstan.

Uranium is chemically extracted from the ore that is concentrated, crushed and finely pulverised to make a yellow paste called yellowcake containing around 75% uranium. This is subsequently transformed into uranium hexa-fluoride, UF_6, which is used for isotopic enrichment.

The fuel can be used in different chemical forms: metal, alloy, oxide, mixed oxide and carbide. The PWR uses uranium oxide, UO_2, the manufacture of which requires complicated technology and very strict quality control. In these reactors, the fuel is placed within a cladding that insulates it from the primary circuit and confines the waste products. PWRs need uranium enriched in U-235 to about 3.5 %. Various enrichment methods are possible [47]:

- UF_6 gas diffusion. The Georges Besse factory at Pierrelatte in France, completed in 1982, produces more than a quarter of enriched uranium worldwide. This method is expensive in terms of investment and operation. It also uses a high amount of energy.
- Gas centrifuging is more commonly used today because of its lower costs. The initial investment is high, but its operational costs are low.
- Other methods are possible and will perhaps be used in the coming decades. One possibility is the separation of an atomic vapour by laser, which makes use of the fact that the ionisation potential of U-235 is different from that of U-238. It is therefore possible by selective ionisation to ionise one of the isotopes but not the other. This procedure is not yet economically viable, and research on the subject has been provisionally abandoned, though it could become viable one day through further technological progress.

The fuel used in PWRs is enriched uranium oxide in the form of pellets, which are cylinders of approximately 1 cm in diameter and 1 cm thick. Each pellet weighs ~ 7 g and liberates, in a PWR, as much energy as about half a tonne of coal. The pellets are contained in tubes some 4 m high called *rods*. These are assembled in bundles placed in the reactor core. The fuel rods remain in the reactor for 3–4 years and are partially renewed every year.

The energy produced in a PWR comes from the fission of U-235 nuclei but also of Pu-239 formed from the U-238 in the fuel. When 100 nuclei of U-235 fission, 55 nuclei of U-238 are transformed into Pu-239. Some of these will fission in return and produce electricity. Around one-third of the energy produced in the reactor comes from the plutonium created.

---

[10] This is some 1000 times more than the average gold content.

A PWR of 900 MW consumes 27.2 tonnes of uranium enriched to 3.2% every year, derived from 150 tonnes of natural uranium [47]. Enrichment represents 31% of the cost of the fuel, and 7% of the total cost of the KWh.

Not all the fissile fuel is burnt. The energy derived from the fuel is called the *combustion rate*. Scientists are trying to increase this rate that would enable more energy to be produced with the same amount of fuel.

The current combustion rate varies from 33,000 to 42,000 MWJ/t, but there are hopes that this can be increased to 60,000 MWJ/t.

4.10   Production of electricity

For reasons of energy and economic independence and security of supply after the first oil crisis, France embarked on a vast programme of nuclear power station construction in order to respond to the growing demand for electricity. Figure 4.3 shows the increase in installed nuclear power in France after 1960. A similar trend occurred at a global level (Figure 4.4) as installed nuclear power increased from 21.3 in 1970 to 357 GWe in 2000 and 381 GWe in 2006. In France, 74.6% of electricity is of nuclear origin (395 out of 517 TWh [5]).

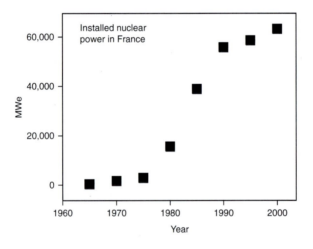

Figure 4.3   Installed nuclear power capacity in France [Source: ELECNUC, CEA, 2001]

At the time of the first oil shock in 1973, France's energy dependence was 77.9%, of which 70% was on oil. In 2000, it was only 49.8% and fell to 43% in 2006 [11].

In France, there are currently 58 PWR nuclear power stations: 34 of 900 MWe, 20 of 1300 MWe and 4 of 1450 MWe. At the beginning of 1999 [1],

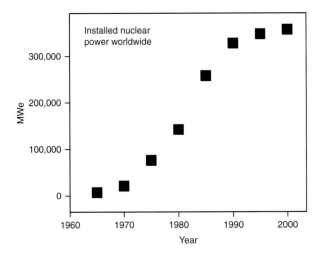

Figure 4.4   Installed nuclear power capacity worldwide [Source: ELECNUC, CEA, 2001]

the installed power was 63 GW compared to 25.2 GW of hydropower and 27.1 GW of thermal power.

The investment cost of a nuclear power station is around 1700 euros/kWe for power stations currently under construction. Availability is around 80–90%, and the life expectancy should be at least 40 or even 60 years. As the installed power is considerable, a major investment is necessary (1.7 billion euros/1000 MWe). This is an obstacle in a liberal economy where investments with quick returns are preferred.

On the other hand, in a country intent on preserving the economic welfare of its population and encouraging its own industry, nuclear energy is a good choice because the price of the electricity produced is low and remains fairly constant throughout the life of the reactor, which should be as long as 40–60 years. This price stability comes from the fact that the price of the fuel is only a small part of the cost of the final kWh, whereas it amounts to some 70% in the case of gas.

The cost of the electricity produced is strongly dependent on the way a country's nuclear power stations are managed [48]. Because it has standardised its reactors, France is particularly competitive and can produce a kWh for a cost of below 3 euro cents. In Germany, it is 62–71% more expensive, according to the rate of discount applied (5% or 10%). In the United States, the cost is on average 30% higher than in France.

The other advantage of producing electricity from nuclear energy is that the price of electricity remains stable over a long period.[11] As the energy is

---

[11] The price can even go down; for instance in the United States, the price per kWh of 95 power stations fell by 24% between 1993 and 1998 [49].

highly concentrated, the fuel can be stored for several years, which avoids price fluctuations experienced for oil.

For nuclear energy, the cost of uranium ore is not the main cost as it is for fossil fuels. A major part of the added value depends on the country that is using it. For instance, if the price of natural uranium increases by a multiple of 10, the price of the electric kWh produced increases by less than 40%. If the same thing happened with gas, it would lead to an increase of 600% (7 times more). The price of gas follows that of oil, and there is every reason to think that it will increase in the future since the peak production of cheap gas will be reached within 20 years.

The price of electricity produced by French nuclear reactors in 1995 was 2.9 euro cents. Of this sum, the fuel represented 0.5 cent, the processing and storing 0.41 cent, the provision for decommissioning 0.15 cent, running costs 0.91 cent, financial charges 0.76 cent and central charges 0.15 cent [49]. The price per kWh today is around 3 euro cents. As electrical consumption in France is about 450 TWh/year, a variation of 0.1 cent in the price corresponds to a total of 450 million euros.

4.11   Rapid neutron reactors

Rapid neutron reactors (RNR) will probably play a more important role in the future. In principle they are simpler as there is no need to slow down the neutrons. As the probability of fission of U-235 and Pu-239 induced by rapid neutrons is weaker than with thermal neutrons,[12] a fuel richer in fissile material has to be used – U-235 or Pu-239 (more than 15%). The neutron flux is also much greater than that in a thermal reactor.[13] The coolant needs to be as transparent as possible to neutrons, which limits the choice. Sodium, mercury or lead can be used.

Some 15 RNRs of output ranging from 12 to 1240 MWe have been built, as well as several prototypes with a lower output. The first nuclear reactor to produce electricity in 1951 was an RNR.

In the 1970s, economists forecast a high growth in demand for electricity in France by the year 2000. To meet this demand, it was necessary to construct not only slow neutron reactors but also rapid neutron reactors. France chose a process in which sodium is the coolant, which was the best compromise choice at the time. The result was the building of Superphoenix, an industrial proto-type also able to generate electricity. This was stopped in 1997 for political reasons, a clear illustration that decisions taken in the energy field are not always rational. In fact, in both economic and scientific terms, it would have been better to allow it to run for another 10 years, which would have enabled further research to have been carried out on the treatment of waste, as well as

---

[12] More than 300 times smaller than for uranium and more than 400 times for plutonium.
[13] 3.5×10^{15} n/cm^2/s in a Superphoenix type RNR as against 2.5×10^{14} n/cm^2/s in a PWR.

generating more electricity.[14] The immediate closure that was imposed cost incomparably more than a programmed closure 10 years later (decommissioning costing almost as much as the construction and commissioning).

A rapid neutron reactor is able to extract in several stages all the energy contained in uranium, whereas a slow neutron reactor can only extract 0.5–1%. For an output of 1000 MWe/year, an RNR only requires 1 tonne of natural uranium, whereas around 2000 tonnes would be needed for a PWR. It also has the advantage of being able to use a large variety of fissile nuclei. An RNR can be operated in three modes to produce energy:

1.   The *fast breeder* mode, in which the reactor produces more plutonium than it consumes.
2.   A mode in which it produces as much plutonium as it consumes.
3.   The *incinerator* mode where the reactor consumes more plutonium than it produces. It can therefore also be used to transmute and destroy other long-lasting radioactive waste.

We are currently in times of energy abundance and slow neutron reactors are sufficient to meet demand. Nevertheless, future sustainable economic development will require us to harness rapid neutron reactors in the long term. Chapter 7 will consider what these future rapid reactors might be like. However, they are unlikely to be economically viable till 2050.

4.12   Nuclear waste

All industrial activity generates waste and the nuclear industry is no exception. In France, some 1 kg of radioactive waste is produced per head of population per year. Out of this, 990 g will decay to a natural level of radioactivity in less than 300 years, but 10 g will remain highly toxic for much longer. It is this highly toxic waste that is the main focus of concern. In comparison, there are 2500 kg of industrial waste per head of population per year of which 100 kg are toxic.

Radioactive waste is made up of matter which cannot be used further, and whose level of radioactivity does not allow it to be directly disposed of in the environment. The most highly active radioactive waste comes mainly from exhausted fuels from nuclear reactors. There are two views on how these should be managed:

1.   The first consists of considering the whole of the exhausted fuel as waste. This is called *open-cycle* treatment. It is the route used up till now by the United States with the aim of reducing the risk of nuclear proliferation.

---

[14] The cores installed could have produced enough electricity to power the city of Lyon for 15 years.

2.   The second consists of separating what is re-usable (uranium and pluto-
      nium) from what is not, with the aim of recycling usable material. This is
      called *closed-cycle* treatment and is the method followed in France.

In all industrial areas, attempts are made to sort and separate waste to
extract what is re-usable from what is not. The final waste is then reduced in
volume as much as possible. The French nuclear industry adopted this philo-
sophy, and the reprocessing of exhausted fuel to extract final waste is fundamental in reaching this goal. During the reprocessing, uranium, plutonium,
fission products and minor compounds are separated by chemical processes. In
France, these operations are carried out at the Cogema factory at La Hague on
the Cherbourg Peninsula. The resulting slight extra cost, as in all waste disposal, is largely compensated for by the fact that the volume of the final waste
is much smaller than that of the exhausted fuel, so the cost of storage will be
reduced. There are also reprocessing plants at Sellafield in the United Kingdom, at Rokashomura in Japan and in Russia.

Figure 4.5 shows the main elements contained in a tonne of exhausted fuel
from a PWR of 1300 MWe after 3 years of cooling [5]. Uranium makes up
95%, plutonium 1%, fission fragments 3.6% and minor actinides (neptunium,
americium, curium) 0.09%. In an open cycle, this represents nearly 1 tonne of
waste, in a closed cycle only 36.5 kg.

A closed-cycle strategy assumes that the reprocessed uranium and plutonium will be used. When there are a number of RNRs, these elements can be
immediately recycled. Without them, the problem is more complex because

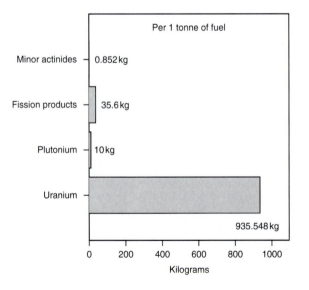

*Figure 4.5   Main elements contained in a tonne of exhausted fuel from a
             1300 PWR after 3 years of cooling [Source: CEA, 2006 [5]]*

plutonium decays, because Pu-241 is transformed through beta radiation with a half-life of only 15 years into Americium-241 that is undesirable in reactors. This means that if plutonium is stored for too long a period this americium has to be separated out. Currently, part of the plutonium produced is used to make *mixed oxide* fuel (MOX) that is a blend of uranium oxides and plutonium. A third of French PWRs can use this fuel and some do. The MOX is however retained after use. There are currently studies to design new architectures for MOX fuel as well as on fuels enabling the recycling of exhausted fuels after successive reprocessing. In all cases, the aim is to reduce the stock of plutonium, but this is a slow process and we will have to await the arrival of rapid reactors to make better use of this resource.

At present, an average of 1200 tonnes of radioactive fuel is produced per year, and 850 tonnes are reprocessed. The reprocessing of fuel and the use of plutonium adds a 1.3% surcharge to the price of the kWh [50].

4.13 Waste disposal

Radioactive waste is classified as short- or long-life waste.[15] Short-life waste is made up of radionuclides whose half-life is less than 30 years. After a period equal to 10 times the half-life, radioactivity will be comparable to that of natural uranium. Long-life waste has a half-life of over 30 years. Radioactivity can be weak, medium or strong, which gives rise to three categories of waste:

1. *A wastes* are short life (half-life under 30 years) and their activity is low or medium (beta and gamma emitters). They make up 90% of the volume of radioactive waste, and 5% of the total radioactivity (15,000 m^3/year in France [49]). These are essentially technological waste from the nuclear industry (gloves, filters, etc.), from users of radioelements (hospitals, laboratories, industry) and research laboratories.
2. *B wastes* are long life. They have weak and moderate activity (alpha, beta and gamma emitters). They represent 9% of the total volume of radioactive waste (1500 m^3/year in France). They mainly come from reprocessing plants (parts of fuel claddings, etc.).
3. *C wastes* contain long life highly radioactive elements (alpha, beta and gamma emitters). They come from the exhausted fuel of nuclear power stations. Included in this category are minor actinides (americium, curium, neptunium) and some fission products (^{129}I, ^{99}Tc, ^{135}Cs). Their volume amounts to 200 m^3/year in France.

In addition to these categories, there is also *very low-level waste* whose activity is of the order of tens of Bq/g:

---

[15] Wastes are only polluting when discarded into the environment.

- Artificial short- and medium-life radioactive elements from the nuclear industry, mainly from the decommissioning of nuclear installations.
- Long-life natural radioactive elements from industries using processes that concentrate the natural radioactivity present in certain ores.

The average volume of waste generated by nuclear power generation was reduced by a factor of 3.8 between 1985 and 1995. Since 1991, liquid and gas effluent wastes have been reduced by a factor of 10.

Highly radioactive, long-life waste, which represents 10 g per head of population per year in France,[16] is the most dangerous. The French parliament passed a law on 30 December 1991 to increase research on this subject over a period of 15 years, under three topics:

1. The separation and conversion/transmutation of radioactive elements.
2. Reversible or irreversible storage in deep geological formations accompanied by experiments in underground laboratories.
3. Treatment and long-term temporary storage on the surface.

It should be noted that there are already solutions for waste management that could be used in an urgent situation, and research is underway to improve these solutions.

Under the first topic, various solutions for the separation of various long-life radionuclides are being studied, in order to store them or *transmute* them in water reactors, rapid neutron reactors and certain dedicated installations such as hybrid systems.

A hybrid system is made up of a sub-critical nuclear reactor into which are injected neutrons produced by an external process, such as a proton accelerator of 1 GeV of high intensity. The protons bombard a target and create neutrons by spallation. Although these hybrid systems are interesting for research, they are unlikely to ever be economically competitive because they are dedicated systems that are different from the rest of the nuclear reactors in service. Waste involved in transmutation consists essentially of minor actinides that only represent less than 0.1% of nuclear waste (Figure 4.5). There will always be some which remains. In such a system, the proton accelerator is very complex and needs to be extremely reliable. Accelerator technology is still far from being fully developed today.

Plutonium is already being burnt in thermal reactors (MOX fuel) and minor actinides could also be slowly burnt. However, rapid neutron reactors are much better incinerators, and can generate electricity at the same time. They would seem to be the best solution for achieving transmutation.

The research under the second topic is aimed at perfecting technologies, preferably reversible, of underground storage. Storage at great depth is the simplest solution for long-life waste, and we have the proof of Oklo on this subject. Unlike at Oklo, these wastes will be sealed in glass. Even in conditions

---

[16] 1 g highly radioactive and 9 g low radioactivity.

where the glass leaches into water, only one ten-millionth of the glass is lost with current technology, which means that in 10,000 years, it would still be 99.9% identical to its original state. At Oklo, the waste was not treated, but simply left in the soil. It would be desirable for the storage to be reversible in order to be able to recover the waste if future technology enabled it to be destroyed or to be reduced in volume. However, storage that is irreversible today may become reversible tomorrow.

The third topic aims at perfecting treatments and methods of long-term storage, until such time as definitive solutions are applied. In the nuclear domain, temporary storage is considered differently from permanent storage. The question that needs to be asked for any particular waste is to know whether it is better to store it provisionally, while hoping to find in the more-or-less-distant future methods enabling radioactive emissions to be strongly reduced, or to find a permanent storage solution for it.

Before temporary or permanent storage, waste must be treated in order to immobilise and confine the radioactivity for as long as possible. This treatment uses a stabilising material that immobilises the waste within a matrix. This is then surrounded by one or several envelopes that provide an additional barrier. The whole package is called a *cylinder*.

In France, the parcels of low activity waste are stored after treatment on the surface by ANDRA in the department of Aube. High- and medium-level long-life waste is temporarily stored there. Parliament will decide on their ultimate destination. For the moment, apart from short-term low- and medium-level waste, the treatment of waste is the subject of research, although there are already solutions in existence. However, the importance of the issue means that research should not be rushed, because there is no great urgency to take a decision.[17]

4.14 Uranium reserves

Uranium is a fairly common element found in rocks and seawater. There is an average concentration of 2.8 ppm (2.8 g/t) in the Earth's crust, sedentary rocks contain less (2 ppm) and granite more (4 ppm). In Australia, there are deposits of 0.5 kg/t, whereas in Canada there are richer deposits (200 kg/t). The concentration in seawater is low (0.003 ppm).

The level of reserves at any one time depends on the price that companies are prepared to pay to recover them. There were considerable stockpiles of uranium in Western countries and the former USSR, which had the effect of lowering prices and putting a brake on new prospects. The price of uranium fell from $42.9/kg in July 1996 to $26.52/kg in 1998 [52]. Since 2004, prices have risen considerably and regularly and reached $130/kg by the end of 2006. Annual global consumption is of the order of 60,000 tonnes, but only 36,000 tonnes come from mined deposits. The rest comes from the decommissioning of nuclear

---

[17] In any case, the waste has to wait to cool down first because radioactivity generates heat.

weapons and the reprocessing of exhausted reactor fuel. The French company, AREVA, controls 20% of the world market in natural uranium.

Known reserves, recoverable at a cost equal to or below $130/kg, are of the order of 4.7 Mt [52]. Adding speculative reserves of around 10 Mt brings the total to 14.7 Mt. At the current rate of consumption, known resources of less than $130/kg would only represent 80 years of reserves for slow neutron reactors. The other reserves that could be exploited would feed thermal reactors for more than 200 years. With RNRs, which extract far more energy from the fuel, there would be sufficient reserves for more than 10,000 years.

The geographical distribution of uranium reserves is very different from that of hydrocarbons. Australia has nearly a quarter of proved reserves, and Kazakhstan and North America each have around 17%. Countries with the most important uranium reserves are shown in Figure 4.6.

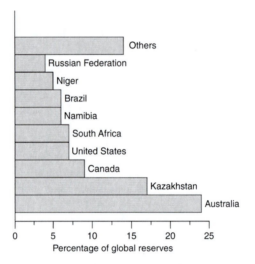

Figure 4.6   Countries with the largest known reserves of uranium [Data from www.world-nuclear.org]

The nuclear power industry is based on the uranium cycle. It is also possible that in the distant future reactors based on thorium could be developed. This element is more abundant than uranium with around 2.5 times more in the earth's crust. Ores that contain high levels of thorium (10–15%) exist.

4.15   Conclusion

Nuclear power is a young energy that has already demonstrated much of its potential. It should play an important part in meeting the growing needs of the world's population. Fears about the increase in the greenhouse effect are a new factor in favour of its wider use, but this is not the only advantage. Its main attraction is its ability to produce electricity at a highly competitive price [26]. In France, which has developed a front-ranking nuclear industry, the low cost of electricity is largely due to nuclear power stations. Gas-fired power stations, on the other hand, rely on imported fuel, which makes up 70% of the price of the kWh.

Nuclear production of electricity in France in 2000 was 395 TWh. Some 87.7 MTOE of oil would have been needed to produce the same amount of electricity, which corresponds to 642.7 million barrels of crude oil. At a price of $25 a barrel that would have cost $16 billion. In addition, overproduction of electricity in France enables the country to export €2.75 billion's worth of electricity per year, and reprocessing brings in around €1.5 billion in foreign currency. In total, nuclear power makes an annual contribution of around €20 billion to the French balance of payments. In 2000, the cost of oil imports in France was €20.9 billion. But that was at a cost of $25 a barrel. At today's prices, it would cost France around €1000 per year per head of the population to generate in oil-fired power stations the amount of electricity produced today by nuclear reactors, which underlines the importance of nuclear power to the French economy. It has improved the balance of payments, since France has little by way of its own energy resources.

At the European level, nuclear represented 15% of total energy consumption in 1997 (212.6 MTOE). France's share of the generation of European electricity was 46.7%. The advantage of nuclear energy in Europe is first and foremost energy independence and reduction of greenhouse gas emissions. The quantity of carbon dioxide emissions avoided by the nuclear process is more or less equivalent to that emitted by 200 million European motor vehicles.

At the global level, nuclear is not an answer to all energy situations. It has no relevance to countries with a low population density, and it is not to be recommended in countries that are politically unstable. On the other hand, it is useful for densely populated areas everywhere.

Worldwide, more people are moving to the towns and cities, and it is estimated that by 2025, 70–75% of the world population will be urbanised, and that the proportion in industrialised countries will even reach 85%. It is difficult to supply energy to all these people without concentrated continuous methods of electricity generation that can be modulated to meet demand.

There remains the problem of nuclear waste, which should not be underestimated. There are already solutions in existence, but no final political choice has yet been made on their adoption. Long-life waste represents, in France, 1 g per head of population per year. The volume of this type of waste resulting from the supply of electricity throughout a whole human life can be contained in a bottle of beer,[18] while a whole lorry would be required to contain the corresponding toxic industrial waste.

---

[18] If the vitrification process is included, this volume should be multiplied by 10.

Chapter 5

Utilisation and storage

Energy is only of value if it can be put to use. For this to happen, equipment is needed to transform the primary energy as quickly as possible into a form that can be directly used by the consumer. The idea of an energy vector that allows energy to be transported over long distances originated at the same time as electricity. But electricity is volatile and cannot be produced just when it is needed, because methods of storing large quantities are not always available and are expensive. The storage of energy is currently a weak point in our energy supply equation and means that certain production methods have to be over-represented.

5.1   Electricity

Electricity is a very convenient energy vector. The number of appliances using it as a source of energy in a modern household is increasing all the time. The electric iron, invented as early as 1880, was revolutionary when compared with a flat iron heated on a stove or embers. To become a market leader, electricity had to wage a fierce battle at the beginning of the twentieth century with other energy sources such as wood, coal, coal gas, candles, etc. It gradually became widely used because of its convenience and economic competitiveness.[1] However, the production and transport of electricity requires major investment, which makes it a very heavily capitalised industry. Further information on this subject is available in Reference 53.

5.1.1   Growth in consumption

Global electrical consumption is increasing faster than that of energy as a whole. Between 1950 and 1990, electricity consumption rose by a factor of 12, while total energy consumption only quadrupled. Between 1950 and 1960, global consumption of electricity was increasing at an annual rate of more than

---

[1] Around 1900, for the same lighting power of equal duration, an incandescent bulb cost 1 F, an oil lamp cost 1.68 F, oil 0.64 F, gas 1.52 F and an electric arc 0.28 F [53].

7%. Since the oil shock, this growth has been weaker. Figure 1.2 shows the growth of domestic electric consumption for France.

Over the past half century, there have been important changes in the utilisation of electricity [53]. In France, in 1950, 76% of electricity was used by industrial sector and 24% by the residential and tertiary sectors. In 2000, industrial sector was only using 34.5% and residential and tertiary 62.2%, nearly half of the latter consumption being used for the heating of air or water. In 2005, 135.8 TWh were consumed by industry (including the steel industry), which was 32% of total consumption, and 272.6 TWh by the residential and tertiary sectors (64.3%) [11]. Urban and rail transport consumed 12 TWh. On the other hand, the share of the electricity used for lighting decreased from 13% in 1969 to 9% in 1995 (it was 60% in 1900) [53].

5.1.2   Production

Electricity can be generated from a number of primary energy sources: fossil fuels, nuclear, renewable energies. In many cases, the production is indirect because it is necessary to pass via an intermediate form of energy to generate electricity, for instance mechanical energy can be transformed into electrical energy using an *alternator*. Alternators have an excellent efficiency (around 99% for nuclear power station alternators), although mechanical energy is generally produced with a fairly low efficiency.

Certain chemical reactions directly produce electricity (*electro-chemical reactions*). Batteries and accumulators are based on this principle. Batteries can only be used a single time, because the reactive components are consumed during the production of electricity. These elements are regenerated in accumulators by reversing the current. Photovoltaic cells produce electricity directly from sunlight.

Batteries enable electrical current to be used in places where there is no electricity. Their cost is judged according to the service rendered and not per kWh, which would be very high. For example, the price per kWh provided by an ordinary dry battery of the LR6 type is around €450. For button cells, the price per kWh can reach €30,000 or more. Additionally, batteries require more energy for their manufacture than they provide at the time of their use. For some batteries, over 100 times more energy is used to manufacture them than they could ever produce.

5.1.3   Alternating or direct current?

For applications using the Joule effect, that is the releasing of heat from circulating current, it does not matter whether the current is direct or alternating. On the other hand, certain uses such as electrolysis or the supply of electronic components requires a direct current. Alternating current was chosen to feed the main distribution grids because transport losses are reduced as the tension is raised. It is easy to change the voltage of alternating current through transformers and adapt it to transportation and distribution requirements, which cannot be done with direct current. However, direct current could possibly bring decisive innovations for decentralised production.

5.1.4   Centralised or decentralised production?

There can be two extreme approaches to the production of electricity. The first is when production is concentrated in certain geographical locations, the other is production at the location where it is used. The first solution, adopted by the developed countries, requires an electrical grid to distribute electricity to consumers. It enables methods of production to be more economic and environmental friendly, since it is easier to invest in low pollution plant. It also enables production methods to be minimised because consumption is smoothed out over a large number of consumers. In decentralised production, the consumers must have production methods that will meet their peak demand, although they will be under-used most of the time.

It is estimated that seven times higher installed power is needed for decentralised than for centralised production.

In regions with high population density, centralised production is the best solution. But for an area of low population density where there is no grid, decentralised production is preferable, because in this case an expensive grid would need to be constructed, whose cost could never be amortised. There are also intermediary situations where there are population concentrations separated by large distances where local grids may be appropriate. In France, local production is viable for isolated locations for which it is not economic to extend the existing grid.

5.1.5   Transport

The transport and distribution of electricity are important because they enable production and consumption to be balanced in the absence of storage. In order to satisfy demand, power stations must be organised to produce enough electricity on the days of the year when demand is at its highest.[2]

Electricity is transported, shared and delivered to the final client by an *electric grid* made up of electric power lines and equipment designed to control the flux. There is a loss of energy from power lines in the form of heat when the electric power is running. To minimise this loss, electricity should be transported at the highest voltage.

Pylons and cables must be capable of resisting wind, storms, frost or lightning. The cost of a power line depends on the relief of the region to be crossed – a line across a plain costing less than one through mountains. The cables must be made of a good conducting material, but copper is not used because it is too expensive. Aluminium alloys that are lighter and more resistant mechanically are preferred.

---

[2] In France, the maximum power called for on the grid was 77.1 GWe on 17 December 2001. On 21 December 1950, it was 6.6 GWe [5]. On 27 January 2006, the instantaneous power demand reached 86.8 GWe at 6.58 p.m.

In France, there are three main types of grid:

1. The transport grid, which carries large quantities of electricity at 400 kV (around 21,000 km in 1997) or 225 kV (around 26,000 km) over long distances.
2. The sharing grid, coupled to the previous one, which serves a smaller region with voltages of 90 kV and 63 kV (around 45,000 km).
3. The distribution network, which goes to the final customers and operates at 21 kV and 5.5 kV (around 600,000 km) then at 400 V (also 600,000 km).

Losses in the power lines, essentially through the Joule effect, amount to around 510% of production. Some materials are superconductors at low temperature, and electricity can pass through them without losing energy through the Joule effect. Unfortunately, the super-conducting effect is only achieved at very low temperatures, which means that energy is required for refrigeration and maintenance at these low temperatures.

5.2 Heat

Unlike electricity, heat cannot be transported over long distances. Heating networks [54] are mainly used for urban heating. At present, in France, there are 380 heating networks in 240 towns heating 2.4 million dwellings or equivalent. In Europe, urban heating is available to 6% of the population or 22 million people.[3]

Water or steam is used to transport the heat. For small networks, the water is kept at a low temperature (110 °C maximum). For large housing estates, super-heated water is often used (between 110 °C and 200 °C under pressure, more than 15 bar at 200 °C). In large networks serving areas with high population density, steam superheated to 200–280 °C under pressure from 5 to 20–25 bar is used. This system can also be used in industrial installations [54].

Many sources of primary energy enable heat to be produced.[4] When coal and heating oil are used, it is important to minimise the emissions of SO_2 and NO_x. Gas is better because it emits less carbon dioxide per thermal kWh produced. Low geothermal energy is well suited to heat production, and around 60 wells have been drilled in the Paris region. The incineration of household waste is used to reduce its volume, but the energy generated can also be recovered.[5] This also avoids the release of methane into the atmosphere. In France, the installed power is 20 GW producing 23.3 TWh of heat.

---

[3] In France, it represents 5% of the heating market, whereas in Finland, it amounts to half.

[4] The distribution of primary energy sources in France in 1997 was coal (21.3%, heating oil (23.4%), gas (28%), renewable (4.2%), incineration of household and industrial waste (20.5%) and others (2.1%) [54].

[5] The calorific value of waste is on average 2440 kWh/t, as against 7000–8000 kWh/t for coal and 11,000 kWh/t for heating oil.

5.3   Co-generation and tri-generation

When electricity is produced by conventional methods, efficiency is of the order of 35% and the remaining 65% is lost in heat. In addition, between 5% and 10% of electricity generated is lost during transport. The principle of *co-generation* is to make use of both heat and electricity. It is a de-centralised method of production that allows efficiencies of 80–90% to be achieved, which reduces emissions of carbon dioxide and other pollutants. Thus, the quantity of carbon dioxide produced by the thermal and electric kWh combined can easily be reduced by two-and-a half times. In a co-generation plant, heat that cannot be transported very far is used on site. The electricity is consumed locally, and the surplus is injected into the grid. The need for heat governs the size of the plant, which, to be viable, must operate at least 4500 hours per year.

A co-generation plant [55] consists of a main engine or motor, an electric generator, a heat pump and a control system. The main machinery may be a steam turbine, a gas turbine, a combined cycle turbine or an internal combustion alternator. Gas is currently the best source of energy for a co-generation plant. In the future, it may be possible to employ fuel cells, Stirling motors or micro-turbines.

Co-generation is attractive for industries that have high requirements of heat and electricity throughout the year. As this heat is usually steam, gas or steam turbines are the most appropriate source. Co-generation can also be used in urban heating networks, especially in countries that need heating during a major part of the year (Nordic countries in particular[6]). Co-generation is also used in the residential and commercial sectors (hospitals, hotels, offices, etc.).

Tri-generation consists of using heat, electricity and coal at the same time. This can be done with a higher efficiency and lower pollution than when the three are generated separately. The energy from co-generation plants is used to generate cold by compression or absorption. Tri-generation enables cold water to be produced to supply air-conditioning networks or industries that need it (breweries, dairies, etc.).

5.4   Transport

The development of modern society means that we all travel more and more, and means of transport use energy. The railways can work on electricity, and if the energy is properly produced, their contribution to the greenhouse effect is low. This is not so true of road [56] and air transport. Road transport depends everywhere on oil for 98% of its energy needs, and airlines exclusively use kerosene.

---

[6] In 1999, Denmark used co-generation to produce half its electricity, the Netherlands 40% and Finland around 35%. At that time, France produced only around 2% of its electricity by this method [55]. The production of electricity by co-generation is around 10% for the whole of Europe.

In 2000, the world consumption of fuel for road transport reached 1550 MTOE of which 600 MTOE was for diesel and 950 MTOE for petrol. Alternative fuels only accounted for 24 MTOE, among them, in reducing importance, liquefied petroleum gas (LPG), ethanol (essentially in Brazil and USA), natural gas (especially in Argentina and Italy), methyl ester from vegetable oil and ethyl *tert*-butyl ether (ETBE) as additives of diesel and petrol [56]. In 2002, 50% of oil products or nearly 1.8 GT of oil were used in transport. If all possible fuels, including biofuels and electricity, are included, the total consumption linked to transport was around 2.4 GTOE in 2005 [91].

In Europe, around three-quarters of the population lives in towns and cities, and 30% of all transport kilometres are urban. Private and utility vehicles represent 98% of the energy consumption of urban transport and more than 10% of the total carbon dioxide emissions of the European Union.

5.5   Energy storage

5.5.1   Hydraulic

The largest quantity of energy stored is in the form of gravitational potential energy achieved by pumping water into higher reservoirs during periods of low consumption. This water can be used to drive turbines when needed. The method is widely used by EDF (Electricité de France).

5.5.2   Batteries

Electricity does not store well and can usually only be stored in small quantities.

Accumulators enable energy to be stored when they are charged, and the energy can be recovered in the form of electric current when they are discharged. They are characterised by their *mass energy density* (the energy that they store by unit of mass), their *power–weight ratio* (the energy that they can deliver by unit of mass and time) and their *cyclability* (ability to be charged and discharged many times).

Lead batteries have been used for many years in motor cars, and will continue to be used with the change to 36 V batteries and hybrid vehicles. They are essential for photovoltaic systems. Their mass densities are around 30–45 Wh/kg, which rules out portable applications.

Nickel–cadmium and nickel–zinc accumulators operate in an alkaline environment. They are more expensive, but their mass energy density is greater (around 50 Wh/kg for the Ni–Cd).

Two new processes were developed in the 1990s, nickel metal–hydride batteries (Ni–MH) whose mass energy density is around 70–80 Wh/kg and lithium batteries in different versions. The Ni–MH batteries are 25% lighter and 30% smaller than Ni–Cd batteries. The recycling of these batteries is also easier because they only contain nickel and not cadmium. Li-ion batteries, widely used in laptop, have mass energy densities in the range of 130–160 Wh/kg and are constantly being improved.[7]

Na/S batteries have also been developed. The largest of them, installed in Tokyo, has a power of 48 MWh. The energy density of this type of battery is around 150 Wh/kg and its specific power is 200 W/kg.

Table 5.1 summarises the properties of some batteries and fuel cells (Zn–air and Al–air) that will be reviewed in Section 5.7.

Table 5.1 *Magnitudes, for selected couples, of tension when empty V (V), of theoretical energy density ρ_{th} (Wh/kg), of actual energy density ρ_E (Wh/kg), of volume energy ρ_V (Wh/l) and of power density ρ_P (W/kg)*

Couple	V	ρ_{th}	ρ_E	ρ_V	ρ_P
Pb-acid	2.15	252	30–45	70–85	200
Li-ion	3.6	631	130–160	260	800
Zn-air	1.4	1050	200–300	30–330	80–100
Al-air	2.7	8140	350–450	350–700	500–600

Source: Reference 57.

5.5.3 Supercondensers

Supercondensers are appliances made up of two electrodes submerged in an electrolyte that can store energy. They are capable of delivering high power during a very short time but cannot store a large quantity of energy. They are complementary to lead batteries that can store energy but release it more slowly. While the superconductor cannot store more than 5–20 Wh/kg, the electro-chemical battery can store 30–160 Wh [90]. A supercondenser can discharge in around 1 s whereas a battery can supply energy for hundreds of hours. The other attraction of the supercondenser is its very large cyclability.

5.5.4 Flywheels

The flywheel, a weighted wheel turning on a central axis, can store energy in the form of kinetic energy. The energy density obtained is around 1–5 Wh/kg. The major part of the kinetic energy occurs at the edge of the flywheel, since that is where the linear velocity is highest. A flywheel that has a moment of inertia I and turning at an angular speed ω has kinetic energy E expressed by

---

[7] Li-ion batteries have no memory effect, can undergo around 300 charge–discharge cycles, and their self-discharge rate is less than twice that of Ni–Cd or Ni–MH batteries. However, they are fragile, need protection circuits, and age whether they are used or not.

the formula $E = \frac{1}{2}I\omega^2$. A flywheel is mechanically coupled to an electro-magnetic converter that enables the reversible transformation of electrical energy into mechanical energy. The energy increases in line with the speed of rotation and the value of the moment of inertia. A good flywheel should conserve its energy for a considerable time and should be able to turn very fast to store the maximum amount of energy. This calls for material with high tensile strength. Since losses are caused by frictional forces, the flywheel is placed in a vacuum to minimise these. Magnetic bearings can also be used to sustain and guide the apparatus.

5.5.5   Compressed air

Compressed air has been used to store energy for a long time in order that the mechanical energy liberated in the release of gas can be exploited. When this occurs in a gas turbine, electricity can be produced. The storage and recovery efficiency is not very high because when air is compressed, it heats up, and the heat produced must be evacuated. When it de-compresses, air cools, and part of the energy of gas compression is lost. The compressors also have very poor energy efficiency.

The first compressed air energy storage plant was built in a salt mine some 25 years ago in Germany. Its available volume is around 300,000 m³ [58]. This plant enables peak electricity demand to be met. It can supply nearly 300 MW of electricity for 2 h. Other large plants have been built in the United States. One of them, commissioned in 1991 in Alabama, compresses air to a pressure of around 70 bar using surplus electricity at low hours of demand and stores it in a salt mine. The compressed air is used at peak demand times in a gas turbine after it is heated using natural gas. Another plant is planned in Ohio using a chalk mine for storage. It should be able to supply 675,000 households for more than 2 days.

5.5.6   Heat storage

While there is much talk of storing electricity, there is less discussion of storing heat, which is however just as important and under-used. Any material has a calorific capacity and can store heat in larger or smaller quantity according to its own properties. A material of calorific capacity C whose temperature increases by ΔT increases its heat by $\Delta Q = C\Delta T$. The calorific capacity C is equal to the product of the mass m of the material by its specific heat c. This last quantity is a measure of an energy divided by the product of a mass and a temperature J/(kg°C). If ρ is the density of the material, which is assumed to be homogeneous and V its volume ($m = \rho V$), the result is $\Delta Q = \rho V C \Delta T$. For a given volume, a material will store more heat as the product ρc is raised. For example, this quantity is 1.02 (MJ/m³°C) for wood, 1.64 for glass, 2.43 for aluminium, 4.19 for water and 7.75 for concrete [59].

Some salt solutions allow much more heat to be stored per unit of volume, as for instance Glauber salt ($Na_2SO_4 \cdot 10H_2O$) that can store 50 times more energy per unit of volume than concrete [59].

While using the calorific capacity of a body is one way to store heat or cold, changing the phase of a body is a good way of storing or releasing heat or cold. Here, the latent heat of the phase change is harnessed. An example of this type of transformation is when a lump of ice is put into a glass of water to cool it. In the case of water, the calorific capacity is 4.18 kJ/kg, the latent heat of melting ice is 330 kJ/kg, and that of evaporation is 2500 kJ/kg. This means that with the energy necessary to boil 1 l of water, 7.6 kg of ice can be melted, and 6 l of water can be raised from 0 °C to 100 °C. Another way of storing heat and transporting it over long distances is to use chemical reactions. Heat can be used to achieve an endothermic reaction with chemical components, the products of which are then transported before the reverse, exothermic, reaction is carried out.

Heat storage helps to smooth the supply of electricity by heating the elements in off-peak periods and storing energy in the form of heat to be released during the day.

5.6 Hydrogen

In the face of the increase in the greenhouse effect and the rising scarcity of fossil fuels, it is important to find a new energy vector that could be used by road and air transport, which currently depend on oil. For France, surface transport consumes around a quarter of all energy (953.7 MTOE in 2004) and this figure is increasing proportionately and more rapidly to the total energy consumption (in 1973, transport energy costs were 32.4 MTOE).

A new energy vector that could be easily transported, stored and distributed can be produced from various sources of primary energy. It must also be easily converted to other forms of energy – mechanical, electrical, chemical or thermal. Hydrogen meets all these conditions [95]. It is the most plentiful element on our planet, existing mainly in the form of water. It has an energy mass density higher than fossil fuels (120 MJ/kg as against 45 MJ/kg for petrol), but its volumetric energy density is lower.[8] Hydrogen is a less polluting fuel since its combustion only produces water.

Hydrogen in its molecular form does not exist in nature, although hydrogen atoms are extremely abundant, whether in the water of lakes, rivers and oceans or in oil and gas. It can be produced from water, but this requires energy and can also increase the greenhouse effect if fossil fuels are used to do this, unless it is possible to capture and store the carbon dioxide produced.

---

[8] One litre of petrol contains energy equivalent to 4.6 l of hydrogen compressed to 700 bar. The unit of volume for hydrogen commonly used is $N\,m^3$ which is 1 m^3 of dry hydrogen at 0 °C under pressure of 760 mm of mercury.

5.6.1   Production

At present, hydrogen is produced from fossil fuels by steam reforming of methane or light hydrocarbons.[9] This reaction takes place at around 800–950 °C in the presence of a nickel-based catalyst. As the reaction is endothermic, energy has to be supplied and around 0.7 J of methane are consumed in reforming a quantity of methane representing 1 J and obtaining around 1.2 J of hydrogen. The reforming results in a mixture of H_2O, H_2, CO and CO_2. Hydrogen can also be made from coal according to the reaction that, if it was complete, would be expressed thus:

$$C + H_2O \rightarrow CO + H_2 \quad \text{and} \quad CO + H_2O \rightarrow CO_2 + H_2 \tag{5.1}$$

The manufacture of hydrogen from fossil fuels always produces carbon dioxide. Centralised production would not contribute to an increase in greenhouse gases if these could be sequestrated. However, it is more economic to use fossil fuels for energy requirements than to make hydrogen for the same purposes.

The other method of manufacturing hydrogen consists of decomposing water:

$$H_2O \rightarrow H_2 + \tfrac{1}{2}O_2 \tag{5.2}$$

This is not possible simply by heating, as the temperature required is far too high (a temperature of over 3000 °C is needed to decompose water). On the other hand, electrolysis of water can produce pure hydrogen in a single step. This requires electricity that may be produced by several methods. In order not to contribute to the greenhouse effect, renewable energies or nuclear energy must be used – or fossil fuels, as long as the carbon dioxide produced in their combustion is sequestrated. Between 4 and 5.2 kWh are necessary to manufacture $1 \, Nm^3$ of hydrogen by electrolysis. To this of course the cost of producing the electricity has to be added. This means that hydrogen produced by electrolysis is around three to four times more expensive than if it is produced from natural gas.

As it is not possible industrially to directly decompose the water molecule by thermolysis, *thermochemical cycles* have been designed to carry out this decomposition in several stages. Many of them allow the process to be carried out at temperatures below 1000 °C but these are still at the research stage.[10]

Even if research carried out on thermochemical cycles using the heat provided by a nuclear reactor operating at very high temperature should succeed,

---

[9] Some 96% of hydrogen today is produced from natural gas.
[10] The one used at ISPRA by Euratom is:

$$6FeCl_2 + 8H_2O \rightarrow 2Fe_3O_4 + 12HCl + 2H_2 (650°C)$$
$$2Fe_3O_4 + 3Cl_2 + 12HCl \rightarrow 6FeCl_3 + 6H_2O + O_2 (200°C)$$
$$6FeCl_3 \rightarrow 6FeCl_2 + 3Cl_2 (420°C)$$

there is very little chance of this being replicable at the industrial level. For reasons of safety and security, a chemical plant is unlikely to be built next to a nuclear reactor.

5.6.2 Transport

Hydrogen can be transported like natural gas, in the gaseous state under pressure, through underground pipelines. Several industrial hydrogen pipelines already exist and experiences using this system have been positive. As with natural gas, regularly spaced recompression stations are necessary to pump the hydrogen.

The calorific power of hydrogen (PCI $= 2.57$ th/N m^3)[11] is about one-third that of gas (PCI $= 8.55$ th/N m^3). Three times less energy per unit of volume of hydrogen can be transported than of natural gas, and its transport costs around 50% more than that of natural gas. An equal volume of gas transports five times less energy than oil and the cost of transport represents around half the price of gas.

One of the problems of hydrogen is that the extreme cold causes hydrogen pipelines to become fragile but this only occurs if the hydrogen is very pure. Impurities in the gas inhibit this phenomenon in the case of steel and thus enable it to be transported in pipelines.

Hydrogen can also be transported and stored in liquid form. The volume mass of liquid hydrogen is 71 g/l at $-253\,°C$ (that of methane is 415 g/l at $-164\,°C$) whereas it is 0.09 g/l (0.72 g/l for methane) in the gaseous state at $0\,°C$. Its boiling point is $-252.8\,°C$ which makes liquefaction difficult. Some 9800 kWh are required to liquefy one tonne of hydrogen. As its energy power is 33,000 kWh/t, this represents nearly one-third of the energy released. This is much more than with natural gas (around 4%).

5.6.3 Storage

The storage capacity of natural gas in France was described in Chapter 2. The same methods can be used for storing hydrogen, which can also be stored in liquid form with low losses.[12]

For transport use, the volume and weight are crucial for the choice of methods of storage. It is now possible to store hydrogen at 700 bar in tanks of composite materials. Some metals and alloys can fix hydrogen reversibly, forming hydrides: as the reaction is exothermic the heat produced must be evacuated. Hydrogen is liberated when the hydride is reheated. These systems are compact but often heavy and expensive. For an equal volume, it is possible to store 2–2.5 times more hydrogen in the form of hydrides than in the form of liquid.

---

[11] 1 thermie $= 1.163$ kWh

[12] A spherical tank of 1000 l loses by evaporation 1% of hydrogen a day only [60]. A 150-litre tank loses 2%. The NASA Kennedy Space Centre tanks that hold 3400 m^3 lose only 0.05% per day.

5.6.4   Uses

Currently, hydrogen is mainly used for industrial purposes. As an energy vector, its main application is in fuel cells producing electricity and heat.

Hydrogen can also be used directly as fuel for combustion engines with minor modifications. Its combustion produces only water but in combination with the nitrogen in the air, nitrogen oxides can also be formed because of the temperature of combustion.

Liquid hydrogen could also be used in air transport to replace kerosene as airline fuel [60]. This would enable pollution to be reduced considerably.[13]

In the past, town gas was made up of about 50% hydrogen and 50% toxic carbon monoxide, and the problems of distributing this gas are well known. In the distant future, when natural gas will be scarce, it could be replaced by hydrogen.

Hydrogen can also be used as a storage method in energy production systems such as wind generators which operate intermittently. It could be produced by electrolysis and used when there is no wind in a fuel cell. The efficiency, however, is low, around 30%.

5.6.5   Dangers

As with any fuel, there are dangers connected with the production, transport, storage and use of hydrogen. However, it diffuses very rapidly, which, outside confined spaces, enables it to rapidly disperse in the atmosphere. As it is very light, it rises very quickly. Its limit of flammability is 4%, close to that of methane (5%).

Although hydrogen has long been associated with danger (the accidents of the Hindenburg and the Challenger), it is very safe if managed properly. The Hindenburg accident was not caused by a hydrogen fire, but by a fire in its envelope caused by an electrostatic discharge in the atmosphere.

Hundreds of kilometres of hydrogen pipelines have worked without problems for more than 70 years. There is much experience of using this gas of which 5 Mt are produced yearly worldwide (955 billion $N m^3$ in France) for chemical and other industrial uses.

5.7   Fuel cells

Fuel cells carry out the reverse of the electrolysis of water [95].

$$H_2 + \tfrac{1}{2}O_2 \rightarrow H_2O \tag{5.3}$$

---

[13] Air traffic contributes 2% of greenhouse gas emissions and 13% of transport-related emissions. Also, it destroys the ozone at high altitude.

Instead of reacting in an explosive manner, as when a stochiometric mixture of hydrogen and oxygen is ignited, the combustion in a fuel cell is slow due to the presence of a catalyst. Its principle was discovered in 1839 by W.R. Grove, but it only came into use in the second half of the twentieth century.

When hydrogen is used as a fuel, only water is produced, so if the hydrogen is manufactured without emissions of greenhouse gases, the fuel cell is a very clean energy conversion system. Other commercial fuels containing hydrogen can also be used: natural gas, LPG, heating oil, petrol, methanol or ethanol. In these cases, the fuel has to be decomposed to extract hydrogen, which is carried out by a reformer. For fossil hydrocarbon fuels, the reformation is carried out at high temperature (600–900°C). In the first stage, steam transforms hydrocarbide into a mixture of hydrogen and carbon monoxide. A second reaction uses water to transform carbon monoxide into hydrogen and carbon dioxide. Some fuel cells are very sensitive to carbon monoxide that is a poison for the catalyst, and in this case a certain threshold must not be passed, which can cause technological problems.

The core of the fuel cell [61, 95] consists of an electrolyte placed between two porous electrodes containing the catalyst. Its general principle is as follows (see Figure 5.1). The hydrogen is directed onto the anode and air containing oxygen onto the cathode. At the anode, the hydrogen is ionised into an electron and an H^+ ion (which chemists call a proton, but which is not the particle discussed in Chapter 4). The nature of the electrolyte is such that it allows the protons to pass but not the electrons. The electrons are forced to circulate in an external circuit creating an electric current. When they reach the cathode, they are captured by the oxygen and produce the O^{2-} ion. This combines with the H^+ protons to form water, H_2O. During the process, electricity, heat and water are produced. The total efficiency of the co-generation system can reach 80%.

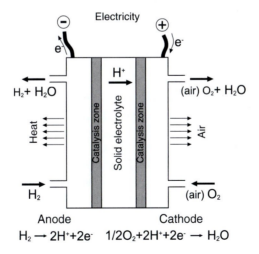

Figure 5.1   Diagram of fuel cell operating with hydrogen and air

The electrical efficiency ranges from 35 to 60% according to the type of fuel cell. Unlike other systems, such as diesel engines or gas turbines, which have a fairly narrow range of operation with an acceptable efficiency, fuel cells operate with a virtually constant efficiency from 40% to 100% of maximum power and can be used between 0% and 100% [61].

Several types of fuel cell have been developed [61], which are at different stages of development:

- The alkaline fuel cell (AFC) uses an alkaline electrolyte (KOH or NaOH).[14] This operates between 120 °C and 250 °C and requires pure hydrogen. This type of fuel cell has been used in the space industry and military applications.
- The polymer electrolyte membrane fuel cell (PEMFC)[15] has a polymer electrolyte and operates between 80 °C and 100 °C. This is the most promising fuel cell for transport applications. It has been the subject of much research, particularly to reduce the costs as the price of the PEMFC technology is currently 7600 €/kW[16] whereas for a car the economic price would need to be below 45 €/kW. There are many potential applications for PEMFC: co-generation for the residential and tertiary sectors (100–300 kWe), transport (50 kWe for cars, 100–120 kWe for buses, more than 200 kWe for rail transport, ships or submarines) and possibly in individual houses with a power of 3–7 kWe.
- The phosphoric acid fuel cell (PAFC) whose electrolyte is phosphoric acid operates at 200 °C. The electrode reactions are the same as for the PEMFC. These fuel cells were the first to be used commercially. Nearly 200 fuel cells of 200 kWe and 200 kWth operating on natural gas, propane or biogas have been installed worldwide since 1995. One is in operation at Chelles in France.
- The molten carbonate fuel cell (MCFC) has an electrolyte that is a mixture of molten carbonates. It functions at 650°C. The electrical and thermal efficiencies are, respectively, 45% and 35%. They are unlikely to have a future because of technical problems linked to the management of molten carbonates and carbon dioxide.
- The solid oxide fuel cell (SOFC) whose electrolyte is a ceramic made of solid oxide operates at 800–1000 °C. The electrical and thermal efficiencies are, respectively, around 50% and 30%. With a turbine, electrical efficiencies of around 60% can even be achieved. This type of fuel cell has a big future in stationary installations. Its advantage is that it operates at high temperature (as with the MCFC) that allows the carbon monoxide to be oxidised and natural gas to be used directly without reformation. The

---

[14] Reaction at the anode: $2H_2 + 4OH^- \rightarrow 4H_2O + 4e^-$. At the cathode: $O_2 + 4e^- + 2H_2O \rightarrow 4OH^-$.

[15] Reaction at the anode: $2H_2 \rightarrow 4H^+ + 4e^-$. At the cathode: $O_2 + 4H^+ + 4e^- \rightarrow 2H_2O$.

[16] Of these €7600, €3800 represent the cost of the electrode–membrane–electrode assembly, and €3050 the cost of the bipolar graphite sheet.

target power would range from 100 kWe to 1 MWe. The SOFC is still at the research stage.

• The direct methanol fuel cell (DMFC) has a solid electrolyte. It operates at a temperature around 120 °C and oxidises the methanol directly into water and carbon dioxide. Its attraction is its potential use in the supply of portable equipment (mini computers, mobile telephones, etc.). The fuel cell is recharged simply by adding methanol.

PEMFC is a very attractive proposition for transport, particularly when hydrogen is used directly. Although the reforming of hydrocarbon compounds can be interim solution, the long-term objective is to use hydrogen. For the time being, these fuel cells are still at the research stage, although a number of prototypes exist. There still remain a number of obstacles to overcome, the most important being price and reliability whatever be the external temperature and vibration. Solutions must also be found to remove the heat from the vehicle but this is not a serious problem because the temperature of the fuel cell is low compared to that of a combustion engine. Great progress has been made on reducing the amount of platinum needed for the catalyst, and in the last few years it has been reduced by a factor of at least 10. However, 3 g/m^2 of platinum are needed and 0.6 m^2 are needed for each kW. For a vehicle of 50 kW, 90 g of platinum are therefore needed. At the beginning of 2000, there were 536 million motor vehicles worldwide. For each to have a fuel cell, 48,000 tonnes of platinum would be needed, which is around 280 times the world annual production.[17]

There are other types of fuel cell for which the energy vector is hydrogen. These are the so-called metal–air batteries that are sometimes called semi-fuel cells. The energy vector here is the metal, which can be changed and recycled when the electrode is consumed. The two couples most studied are Zn–air and Al–air, the characteristics of which are shown in Table 5.1. These batteries, which are still at the research stage, although some of them have already entered production, could be attractive for portable equipment (telephones, mini-computers) or even for electric vehicles.

5.8   Conclusion

Electricity is an energy vector that has proved its superiority despite the large investments necessary. A developed society would have difficulty managing without it. The production of heat and the economies that could be made in producing this final energy is an important area that has probably not received the attention it deserves. In France, energy in the form of heat amounts to double the energy that is in the form of electricity.

---

[17] In 2000, 73% of platinum came from South Africa and 20% from Russia.

Energy storage is clearly a weak point in the energy field and technological breakthroughs are necessary to perfect the products that would satisfy the growing demands for consumers. With current technology, domestic storage of electricity would cost around 15 euro cents per kWh.

Finally, hydrogen will no doubt be one of the energy vectors in the future, but much still needs to be done before it can be used on a large scale (infrastructure, standardisation, safety, security, etc.).

Although a hydrogen-based society is very attractive from the conceptual point of view, intensive use of hydrogen using fuel cells in vehicles is not for the immediate future.

Fuel cells are still far too expensive to be used in transport and many problems remain to be overcome. Since hydrogen does not exist in the natural state, it must be produced, preferably without emitting greenhouse gases. Currently, most of the hydrogen used is derived from natural gas, supplies of which will be exhausted before very long. In the long term, coal could be used, with sequestration of the carbon dioxide emitted. Hydrogen can also be produced by electrolysis, with electricity produced without greenhouse gases (via renewable or nuclear energy) but the cost is around three to four times more expensive than with natural gas. Hydrogen produced by electrolysis will only be competitive with oil when the price of oil is over 75–100 dollars a barrel. The cost would be lower if it was produced from coal. As synthetic fuels can be made from coal, the economic viability of hydrogen in transport applications is still a long way off. In road transport, hybrid vehicles may prove to be the best suited for the use of hydrogen fuel cells in the short and medium term.

It is, however, more probable that hydrogen will be first used for stationary installations using, for example, high temperature fuel cells (of the SOFC type) that have the advantage of being able to reform natural gas without a catalyst.

Chapter 6
Energy and the environment

Both the production and use of energy lead to waste and the emission of polluting gases. Most of these gases are released into the atmosphere, and because of the quantities involved they are implicated in the degradation of the environment. Although energy does make an important contribution to these polluting emissions, other sectors such as industry and agriculture also emit large quantities of gases that are damaging to the environment.

Some pollution is global, affecting the whole of the planet, while some is regional. There are also local sources of pollution that are limited but often more visible. The use of energy also has an impact on health.

Our forebears are often described as protectors of nature, but this is far from the truth; they did in fact have a considerable impact on the environment. The fact that the environment was damaged less in the past is because the human population was less and the standard of living much lower. Neolithic man uprooted wild vegetation, burnt it and used the ashes for fertiliser. They also practised 'slash and burn' techniques that allowed them to sow crops without ploughing. After two harvests, they moved to a new piece of forest and started again. At this rate, it is estimated that a Neolithic family burned around 50 tonnes of wood per year [63]. This corresponds, assuming the wood burnt was green, to an energy expenditure of around 150,000 kWh – much more than a French family uses today. This burning process was not even used to generate energy – it was simply lost to the atmosphere.

We have the impression (although it may not actually be the case) that the environment has been deteriorating since the beginning of the twentieth century. Paradoxically, however, human life expectancy has increased more in last century than in the several preceding millennia. In France, life expectancy increased from 47 years at the beginning of the twentieth century to 80 in 2006, whereas it was only 18 years for Neanderthal man and less than 30 years at the end of the eighteenth century.

6.1 The greenhouse effect

We have earlier described the greenhouse effect and the disturbance caused by gas emissions that have tended to increase this. This section will provide some further information on this subject.

6.1.1   A climate in continual evolution

The earth's environment has changed a great deal since the creation of our planet 4.5 billion years ago. During the Quaternary Era, the earth experienced a succession of glacial and interglacial periods. More recently, there was a small period of glaciation between 1450 and 1880 [64]. Looking at the last million years,[1] the average temperature on earth has been 13 °C, for the last 100,000 years it has been 14 °C and for our millennium 15 °C. We are therefore currently in a relatively warm interglacial period. At the time of the last ice age, 20,000 years ago, the average temperature of the world was 4–5 °C lower than at present [64].[2] However, this small reduction in temperature resulted in major changes compared to our current epoch, as the volume of ice on earth increased considerably. This resulted in a fall in sea level of around 120 m. Even further in the past, the average temperature was higher than at present. In the Cretaceous era, around 100 million years ago, it was probably 6 °C warmer than at present, and the sea level was 300–400 m higher [64]. This hot climate promoted the formation of fossil fuels and more than half of the earth's oil reserves were formed during this era.

6.1.2   The environment

Life has evolved greatly since its first appearance on earth with the appearance and disappearance of new species. A particular case is the dinosaurs of 65 million years ago. The earth's environment is thus not fixed; it develops under the influence of natural phenomena.

The presence of humankind on earth has progressively modified the environment. This is reflected in the disappearance of plant and animal species by the modification of animal habitats, by the development of cultivation and towns, etc. The aim of the human race is to perpetuate itself as long as possible. The human population is now very large, with a high degree of industrialisation. The human impact on the environment is therefore considerable. Although other living beings adapt to the environment as it changes, humans try to adapt the environment to their own needs.

When we speak of preserving the environment, we need a benchmark, but this is not easy to establish because the environment is constantly changing. This benchmark cannot be the state that existed soon after the earth was formed 4.5 billion years ago, and its temperature was sufficiently cool for water vapour in the atmosphere to condense and create the oceans. At that time, humans could not have survived; radioactivity was intense, solar radiation would have been fatal, and there was little oxygen in the atmosphere. On the other hand, the environment of the 1970s is no better than the environment we

---

[1] The first hominid bipeds probably appeared 7–10 million years ago, *Homo habilis* 2.5 million years ago, *Homo erectus* 1.7 million years ago, *Homo sapiens* around 100,000 years ago, and *Homo sapiens sapiens*, from whom we descended, 30,000 years ago. Sedentary agriculture only appeared 8000 years ago.

[2] But there were doubtless considerable disparities; 2 °C lower above the oceans, 10 °C lower at northern latitudes and 2–6°C lower in the tropics.

know today. In the absence of a precise benchmark, we must be pragmatic. We should say that preserving the environment means to keep it in a state in which the human race can continue to develop in harmony. This is therefore not a fixed situation, and we must try to deal with changes occurring to the environment by investigating whether they are harmful or not.

The first 'polluters' that made significant modifications to the earth's environment were unicellular algae. After the earth was formed and the gases of the primitive atmosphere (hydrogen and helium) had escaped, an atmosphere containing carbon dioxide, water, methane and ammonia (NH_3) was created. There was no oxygen and the solar rays striking the earth would have been fatal to life because the ultraviolet rays were not filtered by the atmosphere. After a massive flood that led to the formation of the oceans and trapped part of the carbon dioxide, the atmosphere still did not contain any oxygen. Fortunately, 3.8 billion years ago, unicellular algae carried out the first reactions of photosynthesis. Using solar radiation and carbon dioxide, they released oxygen into the atmosphere.

Over 3.5 billion years, the oxygen concentration increased, enabling animal life to appear on earth, at first in the oceans, to be protected from the harmful rays of the sun. The oxygen concentration progressively increased and an ozone layer was created in the upper atmosphere. This filtered the ultraviolet rays from the sun, allowing life to leave the oceans about 350 million years ago. Therefore, by making considerable modifications to the chemical composition of the atmosphere, the first plants have helped create the conditions for animal and human life that we know today.

So the oxygen that was lacking in the earth's original atmosphere was slowly produced by the process of photosynthesis used by primitive plants as a source of energy. Living beings adapted to the presence of oxygen although it was a highly reactive component and therefore toxic. They therefore developed the process of aerobic respiration. The life cycle was therefore photosynthesis producing oxygen by absorbing carbon dioxide and respiration producing carbon dioxide by consuming oxygen. The creation and consumption of oxygen balanced out to give the current concentration of oxygen in the atmosphere. If there was neither photosynthesis nor respiration, the concentration of oxygen would slowly reduce because the earth is made up of a large number of reducing substances that would gradually absorb the oxygen. It would take about 4 million years to disappear completely. If photosynthesis was stopped, plants would not produce any more oxygen or carbon for the soil. Respiration would continue to produce oxygen, and the rotting of plants on the surface would make the associated carbon disappear. In about 20 years, this carbon would be consumed and the concentration of oxygen would reduce by around 1%.

6.1.3   The greenhouse effect

We have already discussed, in Chapters 1 and 2, the problem of the greenhouse effect. This chapter contains some additional information about it. Further data and analysis can be found in Reference 10.

For hundreds of thousands of years, the concentration of carbon dioxide remained constant between 180 and 270 ppm (parts per million), with slow variations over thousands of years. The lowest values of carbon dioxide were observed during the glacial periods. Up until the last two centuries, it had remained stable around 270 ppm for the 5000–10,000 previous years. Then, from the start of the Industrial Revolution, it has risen steadily to reach 380 ppm today. This increase is due to the burning of fossil fuels. Figure 6.1 illustrates the increase in carbon dioxide emissions in France for the last 200 years. It grew regularly from the beginning of the twentieth century, in correlation with the increase in energy consumption. The first two dips in the graph mark the two world wars: 1914–18 and 1939–45, when the country's economic activity slowed sharply and with it carbon dioxide emissions. In the 1980s, a third dip occurred due to the commissioning of French nuclear power stations.

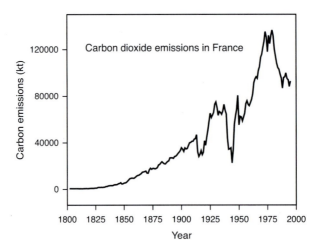

Figure 6.1   Graph of carbon emissions concentration in France from 1800 to 2000 [Source: CEA]

The concentration of methane has doubled in the last 200 years from 0.8 to 1.7 ppm. It remained between 0.6 and 0.7 ppm during the interglacial periods and fell to 0.3 ppm during the coldest ice age periods.

The gases that add to the greenhouse effect do not all have the same level of impact. This depends on the nature of the molecule and the time it is going to remain in the atmosphere before disappearing. To compare the greenhouse effect caused by different gases, the Global Warming Potential index (GWP index) is used [64]. This is the factor by which the emissions of gas X are multiplied to obtain the mass of carbon dioxide that would produce the same impact on the climate in 100 years. This index can produce, for each gas, a carbon equivalent. By definition, GWP (CO_2) = 1. For methane, GWP

(CH$_4$) = 23, which means that 1 kg of methane will have the same impact on climate in 100 years as if 23 kg of carbon dioxide were released. This is why it is better, if methane is not used, to burn it than to release it into the atmosphere. Among other gases, GWP (N$_2$O) = 5700 – 11,900 and GWP (SF$_6$) = 22,200.

Anthropogenic methane is generated by ruminants, rice fields and natural gas leaks. N$_2$O comes from fertiliser. In France, 70% of the emissions in carbon equivalent come from the use of fossil fuels, and agriculture contributed around 20% in carbon equivalent through CH$_4$ and N$_2$O.

Carbon dioxide can also be sequestrated in what are called *carbon sinks*. This is what happens in particular with forestry when what has been cut down is replanted or when forest surface has been extended.[3]

The distribution of the GWP of emissions in France in 1995 after inclusion of the carbon sinks in France was CO$_2$ (68%), CH$_4$ (11%), N$_2$O (19%) and other gases (2%) [10]. In the same year [10], 20% of global carbon dioxide emissions were caused by transport, 15% by the residential and tertiary sector, 30% by industrial sector and 35% by the production and distribution of energy.

Human activities currently emit twice as much carbon dioxide than nature can absorb, around 6 billion tonnes of carbon.[4] This means that a human being emits on average 1 tonne of carbon per year (a French person emits double this). If we wanted to simply stabilise emissions, we could only emit what nature is capable of absorbing, that is 3 billion tonnes of carbon or an average of 500 kg/person/year. In France, we would need to reduce our emissions by a factor of four or five in order to balance the equation. However, 500 kg of carbon per year is the quantity emitted by a small car travelling around 10,000 km. It is therefore clear that it is an extremely difficult objective to reach by the year 2050, the date which has been fixed. Additionally, France contains only around 1% of world population and all countries would have to follow the same restrictions.

Figure 6.1 shows that once in its recent history France reduced its carbon dioxide emissions by a factor of four – when the Second World War was declared. The wartime period brought important changes to living conditions (ration cards, transport restrictions, malnutrition). The reduction only occurred because of the arrival of the wartime economy. This shows how difficult it will be to reduce our emissions by a factor of four in the future.

---

[3] Most of the carbon stored in plants is in the roots rather than the parts above ground. For instance, for a temperate forest, 56.8 t/ha is above ground and 96.2 t/ha below ground [10]. It is even worse in for cultivations (1.9 t/ha above ground and 80 t/ha below ground), or for grasslands (7.3 ha/t above ground and 235.8 t/ha below ground). For tropical forests on the other hand, it is more balanced (120 t/ha above ground and 122.6 t/ha in the soil) [10]. Replacing grasslands with cultivation thus has a negative effect on the carbon balance.

[4] Emissions are either measured in carbon (atomic mass = 12 g) or in CO$_2$ (molecular mass = 44 g). This 1 kg of C = 3.7 kg of CO$_2$.

6.2   Fossil fuels

Every year, many billion tonnes of fossil fuels are extracted. In 2000, 3.5 GTOE of oil, 2.2 GTOE of gas and 2.2 GTOE of coal were consumed. The extraction of these fuels, their refining, transport, storage and use all have major consequences for the environment in terms of pollution and climate change.

The extraction of coal from underground or open cast mines modifies the environment since large quantities of material are moved. Volatile materials are released, such as methane from coal, as well as a large quantity of dust. The leaching of unprotected coal heaps by rain introduces soluble substances into the water tables.

The extraction of oil and gas can lead to subsidence and atmospheric pollution through the release of gaseous or liquid waste. It is estimated that around 5% of gas production is lost. As the greenhouse effect of methane is 25 times that of carbon dioxide, it is as though the whole volume of gas production was released as carbon dioxide into the atmosphere.

Refining leads to the release of toxic products: dust, phenols, ammoniac products, etc. At all stages of the transport and handling of fossil fuels there are losses. Extreme cases are when oil tankers are wrecked, with catastrophic environmental consequences (*Erika, Amoco Cadiz, Exxon Valdez,* etc.).[5] There are also losses during storage.

Coal, oil and gas are rarely pure when they are extracted. They contain sulphur, whose combustion produces sulphur dioxide (SO_2) that is dangerous for health and causes acid rain. The sulphur can be removed from gas before it is distributed. In the case of oil, refining produces light products with low sulphur (petrol and diesel) and heavier products containing sulphur such as heating oil. In the case of coal, the sulphur is not extracted before distribution. The major part of sulphur dioxide produced comes from the use of coal and heavy fuel oil. However, in industrial installations, technology can be put in place to capture a major part of the sulphur dioxide emitted. The technology exists today for constructing and operating clean coal-fired power stations.

Fossil fuels use oxygen from the air when they are burnt. As air contains nitrogen, this can lead under certain circumstances to the production of nitrogen oxides (NO_x).

6.3   Transport

Transport depends almost entirely on oil. Passenger traffic has grown eightfold between 1950 and 1995, and travel by car accounted for 80% of all journeys. The amount of freight transported has also increased. The burning of petrol or

---

[5] Between 1960 and the end of 1999, there were 37 major oil spills including six on the French coastline. Altogether they released 3 million tonnes of crude oil. The illegal purging of tanks adds to this total.

diesel in combustion engines released carbon dioxide that contributes to the greenhouse effect, as well as pollutants (NO_x particles, etc.). The quantity of carbon dioxide released by human activities only represents 3–4% of total emissions; of this 1% originates from deforestation and 2% (around 3 Gt of carbon) comes from the use of fossil fuels. The percentage is small, but a small additional quantity of carbon dioxide causes an important increase in the greenhouse effect. In 1997, transport was responsible for 16% of the total quantity of carbon dioxide emitted by the use of fossil fuels. Looking at the contribution of different types of transport, road traffic represents 62%, air traffic 17%, maritime traffic 7% and other traffic 14% [67]. In 1994, in France, carbon dioxide emissions related to transport amounted to 2.3 tonnes per head of population, representing 38.8% of all emissions.

The pollutants released by transport other than carbon dioxide include:

- Carbon monoxide (CO) that can fix to haemoglobin in blood and prevent it from taking up oxygen.[6]
- Nitrogen oxides (NO and NO_2) referred to as NO_x (nitric oxide (NO) accounts for 90–95% of them).[7]
- Hydrocarbons including methane and non-methanic volatile organic compounds. The methane emitted by transport represents only 0.7% of total methane emissions.[8]
- Ozone (O_3) that results from the photochemical synthesis of hydrocarbons and NO_x.
- Sulphur dioxide (SO_2) arising from the sulphur content of fuel that however is steadily being reduced.[9]
- Particulates (black smoke, particles in suspension) that are caused by the incomplete combustion of fuel. The finest particles can penetrate directly into human lungs and are dangerous for human health. Their long-term effects are not known.

Major progress has been made on the emission of pollutants (CO, NO_x, particulates, hydrocarbons) and their average content has fallen in vehicles from 60 g/km in 1960 to 2.8 g/km in 2000, with a target of 1.2 g/km for 2008. For all French road traffic [70] between 1990 and 2000, CO emissions have reduced by a factor of two, those of volatile organic compounds (apart from

---

[6] Some 59% of carbon monoxide emissions are generated by road transport. They reached 9,380,000 tonnes in 1995, a reduction of 18% on the 1990 figure [12].

[7] The majority (77%) of nitrogen oxides (NO_x) emitted in 1995 (excluding air and sea traffic) came from transport. The quantity was 1,692,000 tonnes. Better energy use through technological advances resulted in a reduction by 8% between 1990 and 1995 [12].

[8] Although the main source of methane emissions ($\approx$ 50%) is livestock, the extraction, transport and distribution of fossil fuels also contributes. Total methane emissions in France in 1995 reached 2,892,000 tonnes [12].

[9] In 1995, France emitted 990,000 tonnes of sulphur dioxide. A major part (43%) came from energy production and transport (13%). France's fleet of nuclear power stations and improvements in energy use have led to a reduction of 70% in sulphur dioxide emissions since 1980.

methane) by 40% and NO_x by 25%. On the other hand, particulate emission increased by 65% between 1990 and 1995, then fell but remained in 2000 some 28% higher than in 1990. This increase was caused by a larger proportion of diesel vehicles. In France, today 70% of new vehicles have diesel engines. Carbon dioxide emissions are constantly increasing because the number of vehicles on the road is increasing, although the new vehicles release less. Between 1990 and 1995, the quantity of carbon dioxide emitted increased by just under 10%.

Freight and utility vehicles contributed 75% of diesel particulate emissions. The black smoke noticed in towns when diesel vehicles accelerate increase with the age of the vehicle. New vehicles are much less polluting thanks to new technology and improved filters. But this is a slow process, because vehicles are only replaced gradually (around 8% per year in France) and the average life of a vehicle is around 7 years. Around a dozen years are needed to renew the major part of the French vehicle fleet. Between 1989 and 2001, the maximum emissions of private vehicles in France were reduced by a factor of eight (from 0.4 to 0.05 g/km) [68]. For sulphur dioxide, which is produced through the burning of the sulphur that is always present in vehicle fuel, coal and domestic or heavy heating oil, desulphurisation processes have been put in place which have reduced its incidence considerably, and the concentration of sulphur dioxide in the air in Paris fell by a factor of 10 between 1965 and 1993 [70].

Pollution caused by vehicles is reducing despite a considerable increase in traffic volume. Between 1970 and 1996, the average pollution caused by motor vehicles fell by a factor of 30. This reduction was particularly marked before the 1990s, because after that technological progress was able to make a rapid reduction in the high levels of pollution originally present. In Paris, in 1993, the concentration of carbon monoxide was only 29% of the 1972 level, that of particulates 25% of the 1960 level, and the quantity of lead had been reduced by a factor of 20 in comparison with 1975 [70]. The concentration of lead in the air in Paris fell from 8 $\mu g/m^3$ in 1978 to 0.2 $\mu g/m^3$ in 1997 [71]. The legal limit imposed by the European Union is currently an average of 2 $\mu g/m^3$/year and it is planned to reduce this to 0.5 $\mu g/m^3$.

Any reduction in fuel consumption leads to a proportionate reduction in the quantity of carbon dioxide emitted per kilometre. Great progress has been made in this area over the last decades, since the reference for European standardisation of vehicles has fallen from 210 g/km of carbon dioxide in 1960 to 185 g/km in 1997. The aim is to reach 140 g/km in 2008 [67].[10]

6.4   Renewable energies

Renewable energies also have an impact on the environment, although it is more limited. Many of them only supply small quantities of energy, so when they are

---

[10] CO_2 emission of 140 g/km corresponds to a fuel consumption of 5.2 l/100 km of diesel and 5.8 l/100 km of petrol (the difference arising from the different densities of the two fuels) [96].

harnessed to provide sufficient energy to satisfy a considerable share of the energy consumption of developed countries their drawbacks become evident.

Industrial hydro has a notable effect on the ecosystem. It can lead to major displacement of the population when dams are built, and more than 1 million people were displaced due to the construction of the Three Gorges Dam in China. Serious accidents are also not uncommon. In France, where there are 89 major barrages, there have been two serious accidents. The first was at Bouzet in 1895, which left 100 people dead, and the second in 1959 at Malpasset caused 421 deaths and laid waste 1000 hectares of agricultural land [72]. Worldwide, there are 35–40,000 barrages higher than 15 m. Between 1959 and 1987, there were 30 accidents at dams causing 18,000 deaths [72]. In 1979, the rupture of the dam at Morvi in India caused some 5000 deaths.

Large numbers of wind generators are unsightly in the landscape and many are needed to produce large quantities of electricity. They can still be noisy, although considerable progress has been made to reduce noise levels. For an equivalent power output, more concrete is needed to fix wind generators in position than to construct a nuclear power station – and this causes a larger emission of greenhouse gases.

In the current state of technology, the manufacture of PV cells requires considerable energy, often generated by fossil fuels. Four to five years are needed to recover this energy, but the PV cell can continue to work for 30 years. A major potential source of pollution are the batteries of independent PV systems, as they contain lead and sometimes cadmium that need to be properly recycled. Much energy is also needed for their manufacture.

Biomass is only of use when what is consumed has been specially planted, otherwise stored energy is simply being burnt as with oil. Problems of land use,[11] soil impoverishment, release of greenhouse gases during the exploitation of biomass[12] are not easy to solve when large areas are involved. Apart from the release of carbon dioxide, which is partly compensated for by what is absorbed during the growth of the plants, the burning of wood releases carbon monoxide, volatile organic compounds and a lot of particulates.

6.5 Nuclear energy

Nuclear energy is frightening because radioactivity is invisible and can be dangerous. However, the nuclear industry is one of the best regulated and supervised industries. Radioactivity is also present in domains other than energy production, such as medicine and industrial verification. Most smoke alarms, for instance, contain radioactive Americium-241,[13] although this

---

[11] If grasslands are turned over to cultivation, carbon dioxide is released, because grassland stores three times more carbon than cultivated soil [10].

[12] N_2O is released during the spraying of fertiliser.

[13] The radioactivity of a smoke detector is around 30,000 Bq.

technology is gradually being superseded. A human body is also radioactive with a radioactivity of around 8000 Bq for a weight of 80 kg.

6.5.1   Units of radioactivity

The bequerel (1 Bq = 1 nuclear decay per second) has already been described in Chapter 4. The number of particles received from an unprotected radio-active source increases according to the proximity of the source.[14] However, the number of subatomic particles received in itself has no bearing on the effect the radiation has on the organism. Thus, billions of neutrinos pass through our bodies practically without any interaction.[15] The important parameter is the energy left in the material by a particle, whether living or inert: this is called the *absorbed dose*. The unit is the *gray* (Gy)[16] and 1 Gy = 1 J/kg. A dose of more than 5 Gy leads to death certainly.

For an equivalent dose absorbed, the effect on a human body varies according to the nature of the radiation and the organ exposed. Dose equivalents are measured in *sieverts* (Sv);[17] 1 Sy = 1 Gy $\times$ Q, where Q is a so-called quality factor that depends on the nature of the radiation. Its value is 1 for electrons and γ rays, 20 for helium nuclei (alpha particles) and between 5 and 20 for neutrons, according to the value of their kinetic energy. The sievert is a large unit, a dose of 10 Sv certainly causes the death of a human being.

The relation between the radioactivity of a source (its number of becquerels) and the absorbed equivalent dose is complex. It depends on various parameters and on the nature of the exposure (internal or external). For instance, adults receive a dose of 1 mSv if they swallow 3500 Bq of Radium-236 – or if they inhale 60 Bq.

The legal limit that a person can receive on top of natural radioactivity is 1 mSv/year. For workers in the nuclear industry, the limit is 20 mSv/year averaged over 5 years.

6.5.2   Natural and artificial radioactivity

Any radiation has exactly the same properties whether its origins are natural or artificial. A French person receives on average a dose of 3.5 mSv/year, of which 2.45[18] is natural and 1.05 is artificial [74]. The breakdown is given in Table 6.1,

---

[14] It is inversely proportional to the square of the distance, so that if the distance is halved, four times more particles are received.

[15] Of the 100,000 billions of neutrinos passing through the earth, only one is stopped.

[16] The old unit was the rad. 1 Gy = 100 rad.

[17] The old unit was rem. 1 Sv = 100 rem.

[18] In some parts of the world, natural radiation can reach 10–50 mSv without any impact on human health. In the Creuse department of Central France, some people receive around 100 mSv/year.

but the data given should be taken as round figures. Doses received by any individual can vary according to place of residence,[19] lifestyle,[20] health,[21] etc. In general, natural radioactivity accounts for 70% of the radiation dose received. The remaining 30% of artificial radioactivity come mainly from medical treatment. Nuclear energy only accounts for 0.1% or 0.004 mSv/year. The contribution of nuclear weapons testing and the Chernobyl accident total 0.02 mSv/year.

Table 6.1   *Estimated radioactive doses received by a person in France [74]*

Origin	Dose (mSv)	%
Natural radioactivity		
Earth's crust	0.42	12
Cosmic radiation	0.37	10.5
Internal (human body, water, food)	0.37	10.5
Radon (air)	1.3	37
Total natural radioactivity	*2.45*	*70*
Artificial radioactivity		
Medicine	1	28.5
Industry	0.035	1
Military nuclear	0.01	0.4
Nuclear energy	0.004	0.1
Total artificial radioactivity	*1.05*	*30*
Total radioactivity	3.5	100

Radon makes the biggest contribution to natural radioactivity levels. Radon is an inert, radioactive gas with a short half-life, whose descendants are also radioactive. Humans breathe it in all the time, especially in badly ventilated buildings. The International Commission on Radiological Protection (ICRP) advises that radioactivity of 66 Bq/m^3 within a dwelling corresponds to 1 mSv/year. In France, 0.5% of dwellings have radon in excess of 1000 Bq/m^3, corresponding to a dose of 15 mSv/year [76].

---

[19] Radioactivity of granitic soil is around 8000 Bq/kg while that of sedimentary soil is around 400 Bq/kg.

[20] A return flight Paris–New York gives a dose of around 0.06 mSv, a week's stay in the mountains at 1500 m 0.01 mSv. In France, a baby given a bottle prepared with mineral water may receive a dose of 0.35 mSv from the water – one-third of what is permitted (1 mSv).

[21] The dose received during a chest X-ray is 0.07 mSv, a dental X-ray 0.1–0.2 mSv, a cardiac angiogram 6.8 mSv and a brain scan 50 mSv (50 times the permitted limit). Radiation doses in cancer treatment range from 20 to 50 Sv.

The major part of artificial radioactivity comes from medical examinations, which is why attempts are constantly being made to reduce the dose while maintaining the level of information or treatment.

Doses below 100–200 mSv/year have not been shown to have an effect on health, and in particular do not cause excess cancers or leukaemias. Although it is known that high doses of radiation can cause cancers, there remains considerable uncertainty about the effect of low doses. As a safety precaution, the risks are assessed by linear extrapolation of the data provided from Hiroshima and Nagasaki, from patients having undergone radiotherapy, and from animal experiments. This extrapolation method almost certainly overestimates the effects, because it is very likely that the self-repair properties of DNA mean that there is no measurable damage below a certain threshold of exposure.[22]

As is clear from Table 6.1, the major part of the radioactive dose received by humans comes from natural sources or medical examinations. The contribution of nuclear energy is negligible. But there are occasional incidents and accidents, resulting mainly from human error, hence the importance accorded to the human factor in safety studies.

The effects of radiation on living tissue are usually described as *deterministic* or *stochastic (random)*. Deterministic effects, which appear above a threshold radiation dose for the irradiated tissue or organ, lead to the massive destruction of cells of the organism. The damage increases with the dose received, and it is seen after a relatively short time after irradiation (some hours to some days). Stochastic effects result from the modification of cell DNA. The genetic make-up is disturbed and may lead to cancers some time later (some years to some decades). This will be examined in Section 6.6.

External irradiation by a radioactive source ceases when the source is removed. *Internal irradiation*, which occurs as a result of contamination by absorption of a radioactive substance, is much more dangerous, because it continues after the radioactive product is eliminated from the body. The speed of elimination depends on the interaction of two phenomena – the metabolism and the half-life of the radioactive substance. The expression *biological half-life* is used to define the time required for an organism to eliminate half of the ingested radioactive substance. This is different from the half-life of the nucleus. For instance, with Caesium-137, the biological half-life is 100 days, whereas its ordinary half-life is 30 years.

6.5.3 Incidents and accidents

The nuclear industry is highly supervised with a rigour that would be appropriate for other dangerous industries to adopt. Since 1991, incidents and accidents have been classified according to the International Nuclear Event Scale

---

[22] The following illustration may be relevant. Drinking two 25 cl glasses of wine per day for a year is not dangerous for health and may even provide certain benefits, but drinking, for only one month of the year, 6 l of wine per day is certainly harmful – although the annual volume consumed is the same in both cases.

(INES) that has 8 levels of gravity. This type of classification was introduced in France in 1987. Level 0 refers to deviations, with no safety significance. Level 1 is a simple anomaly. Level 7 is a major accident. The need to classify the slightest incident is good, because it compels the operator to work rigorously. The history of incidents and accidents is an excellent bank of experience that can be used to improve the safety of nuclear installations on an ongoing basis.[23]

Three serious accidents have occurred in nuclear power stations to date:

1. Windscale, 7–12 October 1957, on a graphite–gas reactor. The accident was Level 5 on the INES scale (accident with limited off-site release). Iodine-131 (740 terabecquerels, or 1000 times less than at the Chernobyl accident) was externally released.

2. Three Mile Island in the United States, on 28 March 1979, on a PWR of 900 MWe. This accident occurred as a result of an accumulation of material failures and human errors. It was classified as Level 5 on the INES scale. There was no contamination of the local environment, despite a partial meltdown of the reactor core, due to the good design of the shell. Analysis of this accident led to considerable progress in safety issues.

3. Chernobyl, which occurred on 26 April 1986, is the most serious accident to have taken place until now. It was classified Level 7 (major accident) on the INES scale. The type of reactor involved (**RBMK**) is inherently unstable in certain operating ranges and does not possess a containment shell as western reactors do. Safety directives were violated by the voluntary blocking of security systems during a safety test that had not been properly prepared. Around 10% of the radioactivity in the core was released.

Chernobyl is an example of what not to do in nuclear energy. The management of the aftermath of the accident was also badly carried out, which led to health consequences more serious than they need have been. However, as is shown in the UNSCEAR[24] reports of 2000 and 2001, which are summarised in Reference 77, the health consequences have been less than has been reported by the media.[25]

There may also be psychological problems linked to the fear of having been irradiated and to be forcibly moved, which can lead to serious symptoms of anxiety and depression and psychosomatic illness. These were observed in the

---

[23] For PWRs, there is now the equivalent experience of more than 10,000 years of operation.

[24] United Nations Scientific Committee on the Effects of Atomic Radiation.

[25] Two hundred people were severely irradiated and 32 died in the three months following the accident (out of 500–600 on the site). Between 1986 and 1990, 600,000 'liquidators' worked on the clean-up of the site. Using the data from Hiroshima and Nagasaki, one would expect some 3000 of them (0.5%) to die sooner or later from the effects of radiation. After 15 years of medical follow-up, an excess of thyroid cancer was found in children, but there does not appear to have been a general increase in case of cancer and leukaemia in adults.

populations close to Chernobyl and Three Mile Island,[26] and in patients followed up at the Curie Institute.

During the Chernobyl accident, around 12 kg of uranium were released into the atmosphere [78]. This is 350 times less than the 4.2 tonnes released during atmospheric nuclear tests carried out by the nuclear powers [78].

In France, the most serious accident occurred on 13 March 1980 at Saint-Laurent-les-Eaux, but there were no human casualties. It was classified Level 4 (accident with minor off-site impact).[27]

6.6   Radioactivity and organic life

Living organisms are composed of cells, which co-operate amongst themselves to make tissue that enables the organism to function. Each cell contains DNA molecules which contain all the information necessary for the proper functioning and reproduction of the cell. Modification of the DNA can have dramatic consequences for the cell. Changes do occur frequently, but fortunately powerful repair mechanisms exist which prevent most modifications that could have serious consequences for the offspring cells of the damaged cell.

In our environment, oxygen is paradoxically the most toxic substance for cell DNA. Its reactivity, which is indispensable for life, leads to the formation of free radicals that can cause cell damage during breathing. Oxygen damage during respiration and metabolism causes around 10,000 DNA lesions per cell per day [79]. Thus, the DNA of each cell undergoes an impressive number of changes every day. Fortunately, there is a highly efficient surveillance system that reacts to all changes in the DNA and deploys enzyme mechanisms to repair them. In the vast majority of cases, the repair is perfect and everything functions as before. In some rare cases, the repair is done badly and then one of two situations occurs. In the first, the cell activates suicide genes and programmes its own destruction (cell death). In the second, an irreversible mutation occurs which is written into the genome and transmitted when the cell divides. This mutation is the first step in the development of a cancer but is not on its own sufficient for cancer to develop. Other genetic and environmental factors are needed to enable the multiplication of the mutated cells. This is one explanation for the long delay before the possible appearance of a cancer.

The simple fact of living means that a human being undergoes tens of thousands of DNA alterations in every cell daily. Interaction with the environment increases this number. For instance, an hour's sunbathing on the beach causes between 60,000 and 80,000 alterations per cell, 5 cigarettes per day between 25 and 50 alterations. Natural radioactivity only causes 2 alterations per cell per year [79]. The contribution of the nuclear power industry, which

---

[26] Among the liquidators at Chernobyl, a higher rate of suicide than in the general population was observed.

[27] A WHO study covering the period 1969–86 listed the average number of mortal accidents linked to the production of energy every year. They were 200 for coal and hydro, 130 for oil, 80 for gas and 3 for nuclear [79].

only contributes an average of a few thousands of an mSV, is entirely negligible. For workers in a radioactive environment, the dose received is generally equivalent to background radioactivity.

Some individuals are hypersensitive to the effects of ionising radiation, others to ultraviolet light. This may be due to a spontaneous chromosomal instability or to a major defect in DNA repair enzymes [80]. The probability of developing skin cancer after exposure to sunlight is thus very high in people who are highly sensitive to ultraviolet rays, and can happen very early in life (4–5 years). Similarly, people sensitive to ionising radiation can develop cancer before the age of 20.

Except in extraordinary situations, the use of nuclear energy has no notable effect on human health. It is much lower than food and tobacco that are responsible for 60% of all cancers [79].[28]

6.7   Conclusion

Any activity can cause environmental problems. The production, transport, storage and use of energy are no exception. Once there is a need for large amount of energy, as in the developed countries, it is difficult to satisfy it without making some impact on the environment. In addition to local and regional pollution, we now have to take into account global pollution linked to gases contributing to global warming. Of the gases produced during energy generation or use, those that are reactive and therefore toxic disperse quickly but cause severe local pollution problems. Those that are not toxic, like carbon dioxide, survive for a long time in the atmosphere.

While centralised generation using fossil fuels can reduce pollution by unit of electricity generated – or even dispense with it altogether one day if the carbon dioxide can be successfully sequestered – transport remains a major problem as a non-localised source of pollution.

The main problem of climate change caused by the emission of greenhouse gases during the burning of fossil fuels or other human activities is that it is happening very rapidly compared to the natural timescale. Major climate changes like the ice ages lasted 80,000 years, and the warmer periods, although epochs of faster change, several thousand years [65]. Nature is therefore likely to have much difficulty in adapting quickly to human-induced climate changes.

In the face of these environmental problems, nuclear energy, if rigorously managed and regulated,[29] is a good solution for producing the large amounts of energy necessary for densely populated regions with high energy requirements. On the other hand, for regions with low population density and modest energy requirement, renewable energies are by far the best solution.

---

[28] The risk of premature death per 100,000 persons is estimated at 21,900 for tobacco, including all causes (8800 for cancer). It is 1600 for road accidents [81].

[29] Chernobyl made a big impression on public opinion towards nuclear energy. Just because your neighbour has a badly maintained car and does not respect the highway code, it does not mean that you should not drive your own car.

Chapter 7
Future prospects

The production and use of energy should increasingly be integrated into a sustainable development plan where resources are better exploited and waste and emissions are minimised. The notion of *sustainable development*, introduced in 1986 by the Brundtland Commission, is defined as development that meets the needs of the present generation without compromising the ability of future generations to meet their own needs. In this final chapter, a number of future prospects in this perspective are examined.

For a very long time, human beings used little energy. Four hundred thousand years ago, they probably used around 0.4 kg oil equivalent (KOE) per head per day to meet food and survival needs (fire). Since the industrial revolution, human energy needs have sharply increased, and, for example, in the 1990s an American consumed some 21 KOE/day [87].

It is interesting to compare the human energy system with human's current energy needs [1]. The energy needed for a human's base metabolism develops over the course of life. At birth, in a developed country, this energy is on average 2.3 W/kg, reaching 2.7 W/kg 3–6 months later. It then reduces and stabilises around the age of 18 for about 40 years at around 1.1–1.4 W/kg. Beyond that age, it falls to just under 1 W [1]. Nearly half this energy fuels the brain and nearly a quarter the liver. A woman's pregnancy calls for additional energy of around 90 kW [1]. For comparison, 90 kWh is the energy contained in 10 l of petrol, sufficient to drive a car for some hundred kilometres.

Food produces energy of around 2.7 kWh/day. A French person's electricity consumption is around 20 kWh/day, which with other forms of energy leads to a total consumption of around 50 kWh/day.[1] It is as though modern man has about 20 slaves to help him throughout his life, but this is probably an underestimate, since many of the appliances we use were made on the other side of the world. The energy for their manufacture was consumed there and the associated waste and pollution were generated there. The energy for transport must be added, which in modern society has huge energy requirements. For example, the four engines of a Boeing 747 in flight have a

---

[1] These estimates were calculated from the fact that the total electricity consumed in France in 2000 was 88 MTOE out of the total consumption of 215.7 MTOE.

power of 60 MW, which corresponds, for a flight of 10 h, to a consumption of 600 MWh [1].

7.1   Fossil fuels

Fossil fuels will still be used long into the future. Further progress needs to be made in oil recovery. To extract rates of more than 50%, techniques of drilling in all spatial directions must be improved and better imagery must be developed to understand the conformation of oil wells, and fluids for making the oil less viscous [82].

The other challenge is to be able to capture and store the carbon dioxide produced in power stations. Experiments are already under way [83]. In Norway, 1 Mt of carbon dioxide is injected annually from the Sleiper offshore gas platform in the North Sea into a saline aquifer 1000 m deep. In Canada, some 5000 tonnes of carbon dioxide from a coal gasification plant are injected into a disused oil field located 300 km away. Other projects propose the burying of carbon dioxide at great depths in the ocean, but that could have an impact on the acidity of the surrounding sea.

The sequestration of carbon dioxide in deep, underground geological reservoirs would enable it to be stored for tens of thousands of years, but potential dangers need to be taken into account, such as an unexpected large-scale release of the gas.[2]

Carbon dioxide sequestration is the most expensive step: the cost of capturing and compressing the gas is estimated at between 30 and 60 €/t [97]. Transport costs around 3.5 €/t per 100 km. Injection and storage would add 20 €/t for a quantity of 1 Mt/year but might fall to 7 €/t for sites storing 10 Mt/year. Carbon dioxide capture uses energy that reduces the efficiency of energy production by an equivalent amount. Consequently, more carbon dioxide is released for the same amount of electricity produced. Various routes have been proposed for capturing carbon dioxide:

- The first is at the post-combustion stage, which would enable it to be applied to current power stations. The carbon dioxide has to be separated by a scrubber from the other exhaust gases.
- The second consists in employing oxy-combustion, which involves combusting coal in an enriched oxygen environment using pure oxygen extracted from the air (which requires energy). The exhaust gases then contain 90% of carbon dioxide that is easier to separate.
- The final route is capturing in pre-combustion where the idea is to capture the carbon dioxide before combustion, when the fuel is manufactured. For this, the initial fuel is reformed with steam (steam reforming) or oxygen

---

[2] The carbon dioxide released by the volcanic lake of Nyos in Cameroon in 1986 caused 1600 deaths.

(partial oxidation) to produce a synthetic gas (blend of hydrogen and carbon monoxide).

7.2 Renewable energies

The most important objective for renewable energy sources is to make them economically competitive (when they are not so already) and deal with their intermittent nature to be able to draw on energy at all times. By 2020, solar photovoltaic energy will most probably supply 1 billion people. Studies are under way on the feasibility of producing large quantities of energy in the deserts (5% of their surface could generate enough energy to supply the entire planet). The short- and medium-term objective for renewables is not to replace fossil fuels but to substitute for them where possible so as to reduce their consumption. It is also important that a country should make the best use of the renewable resources available to it. There is no universal solution and each case must be looked at on its own merits. Solutions for a new house are not the same as for a renovated property and the technologies to be used will depend on the climate and the region.

Marine energy was mentioned in Chapter 3 on renewable energies. Off-shore wind farms are already a reality, but there are other possibilities that could be exploited in the future.

Wave energy (around 1 W/m^2, 45 kW/m of coast) is highly diluted and not yet economically viable. France has a good potential compared with some other countries. On the North Atlantic coast, which is particularly suitable, wave energy could generate electricity for a current price of around 8 euro cents per kWh.

The thermal energy of the oceans is potentially 100 times higher than that of tides or waves (it is estimated at 10^{13} W) [30]. The principle is to use the temperature difference between the ocean surface (25–30 °C in the Tropics) and the water at great depths (7 °C at a depth of 600 m for example). The difference must be more than 20 °C for it to be worth exploiting, and efficiency is low (2%).

Preliminary studies are also being made into how energy could be obtained from osmotic pressure arising from the difference in ionic concentration between seawater and fresh water.[3] This property is already exploited by nature because it drives the thermohaline circulation that is the second motor of the oceanic currents. With fresh water, osmotic pressure is much higher.

Heat pumps, which have existed for many years, will no doubt be further improved. They can be considered as a renewable energy source insofar as a large part of the energy they provide comes from renewable resources. They

---

[3] Just as two bodies with different temperatures equalise when they are placed in contact, two solutions of different concentrations placed in contact have the tendency to equalise their levels of dilution. The variable called chemical potential plays the role that temperature plays in the first case.

work by extracting heat from cold and low temperature sources and producing a fluid sufficiently warm to be used to heat a house or a building, for instance. As the spontaneous transfer of heat from a cold source to a hot source is not possible, outside work is needed to do it, and this is exactly what a heat pump accomplishes. This type of device can be considered as an energy amplifier since by using 1 kWh of energy, the heat pump can generate 3–4 kWh of heat. The difference between what is supplied and what is consumed is taken from the external medium that may be air, water or soil. In the latter case, it is called a geothermal heat pump.

Other more futuristic ideas being examined include the collection of solar energy in space. Orbiting solar power stations were first proposed in 1968. The idea is to place a solar panel of several km^2 in geostationary orbit at 36,000 km altitude. In space, solar radiation is not attenuated by the Earth's atmosphere, and the panels can be manoeuvred to face the sun. Some eight times more solar radiation would be received per surface unit. Microwaves, electromagnetic radiation similar to that used in a microwave oven or a mobile telephone, could be used to transmit this energy to the Earth, with an efficiency of around 50%. However the energy density carried by the microwaves should be sufficiently weak so as not to cause danger to living beings, which implies the need for considerable areas of radio batteries to store the energy.

7.3   Nuclear power in the future

Nuclear energy produces electricity without increasing the greenhouse effect. It is highly concentrated since 1 g of fissile matter releases around 1 MWJ, or 24 MWh. The reactors of the future, designed in a sustainable development perspective, will need to satisfy the following five conditions:

1. They must be competitive in their generating cost compared to other energy sources. The technology chosen must provide as short as possible return on investment. They must be able to operate for a long time, typically 60 years, and they should be easy and cheap to maintain.
2. Progress in improving safety in the nuclear industry is regular and constant, so they will be safer.
3. They will need to extract the maximum amount of energy from the fuel.
4. They will minimise the production of waste and will be capable of burning part of the spent fuel of the previous generation of reactors.
5. They will need to minimise the risks of proliferation.

Bearing in mind these constraints – and especially those relating to resources and their optimal use – fast neutron reactors are the best solution. Like slow neutron reactors, they may use different technologies. An example of this type of reactor is the Superphoenix developed in France but abandoned for political reasons. Discussions and studies are currently under way at the international level within the framework of the Generation IV International Forum (GIF) [98].

Among the possible solutions, we will briefly describe one to illustrate the problems posed.

The first thing to be done is to increase the efficiency of electricity production. To do this, the thermodynamic efficiency needs to be increased by raising the operating temperature of the reactor. At present, due to Carnot's theorem, the thermodynamic efficiency of a PWR is 33%, and 2 kWh of heat is released into the environment for every 1 kWh_e produced.[4] To increase the efficiency, the reactor must operate at a *high temperature*.

Unlike thermal neutron reactors that only use some 0.5–1% of the energy contained in the uranium, fast reactors can extract more than 100 times. Existing uranium reserves would supply nuclear energy produced by them for more than 10,000 years. This more efficient use of the fuel also reduces the amount of waste, partly because the natural uranium is better used and partly because some of the minor actinides are burnt in the reactor. However, it is not possible to extract all the energy in the fuel at one time. The fuel must be recycled, that is *reprocessed*. This takes us on to *fast neutron reactors*.

The coolant in a fast neutron reactor can be a gas (He or CO_2), sodium or lead–bismuth, which absorbs little or none of the neutrons. The simplest architecture is to function in direct cycle (a single circuit). At present PWRs have two circuits: the primary, in contact with the fuel core sheath, and the secondary circuit. Boiling water reactors, which have only one circuit, seem, other things being equal, to have an advantage from their direct-cycle architecture (no steam generator, for instance). Sodium fast reactors are more complex, with three circuits, two of sodium and one of water. A fast direct-cycle reactor using gas as a coolant is an interesting solution. Helium would be a good choice because of its chemical and nuclear inertia, but its low density excludes specific high energies.

The other advantage in using a gas as a coolant is the possibility of developing turbines based on the combined-cycle gas turbines that are an important element in the competitiveness of gas (efficiency above 55%).

The use of high-temperature fast reactors implies serious constraints on the types of fuel used and on the construction materials. The integrated fast neutron flux will also be important. Extremely strong materials will be needed. The fuels will need to be refractory, high-yielding, resistant to radiation and capable of being reprocessed.

To show how well fast neutron reactors fit into a programme of sustainable development, let us take the example of a PWR generating 1400 MW recently commissioned in France. After 40 years' operation, there will be some 7000 tonnes of depleted uranium from the enrichment necessary to make fuel, 1000 tonnes of uranium for reprocessing and 11 tonnes of plutonium. These nuclear materials

---

[4] A fast neutron reactor like Superphoenix operated at a higher temperature and its efficiency was higher, around 45%.

could be used as fuel in fast reactors to provide an equivalent amount of electricity for more than 5000 years instead of 40 years.

To develop nuclear energy at the industrial level, two cycles are possible: uranium and thorium. The uranium cycle was chosen. The fissile nuclei are Uranium-235, which exists in the natural state, and Plutonium-239 and Pu-241, which are artificial. The uranium cycle can be started easily. In the thorium cycle it is U-233, with a half-life of 160,000 years, which is fissile, and it has to be synthesised from Th-232, which is fertile. The thorium cycle is interesting as it produces shorter-lived waste. However, the reprocessing of the waste is very complex because it contains radioactive isotopes emitting gamma rays with an energy of several MeV, which must be protected against. Thorium is more abundant than uranium in the Earth's crust and could be used once uranium deposits are exhausted. Uranium and thorium together represent energy reserves for tens of thousands of years at a competitive cost. Thorium is more readily available than uranium for manufacturing a nuclear weapon but once made it can be more easily detected.

7.4   Energy in the home

Around 43% of the final energy produced in France in 2004 was used in homes for heating, air conditioning, the production of domestic hot water and various electrical appliances [90]. Energy consumption in the domestic-tertiary sector has greatly increased over the years, from only 28% of final energy in 1949. Around one-third of the energy is consumed in tertiary sector buildings and two-thirds in individual homes.

Individual homes are thus a major contributor to energy consumption – and also to emissions, since they contribute 21% of all greenhouse gases in France, corresponding to around 550 kg of carbon per person per year.

In France, there are around 30 million dwellings, and 400,000 new buildings were completed in 2005. Buildings are only renewed over about a century – a long time compared to the constraints necessary to fight climate change. The renovation of old housing stock is thus of prime importance, and this renovation needs to be done without moving the inhabitants out.

Buildings built before 1975 had poor energy characteristics. On average, 328 $kWh/m^2/year$ was needed for heating and 36 $kWh/m^2/year$ for hot water. New regulations have reduced these figures. Thus, from 2000, new buildings should only need between 80 and 100 $kWh/m^2/year$ for heating and 40 $kWh/m^2/year$ for hot water. The target today should be 50 $kWh/m^2/year$ for heating and 10 $kWh/m^2/year$ for hot water. Despite all the improvements in construction methods, the past weighs heavily and the average energy consumption in French homes today is 210 $kWh/m^2/year$ for heating and 37.5 kWh/m^2 for hot water [90].

Homes are also closely linked to transport. There is no point in having a low-energy home if you need to travel 100 km a day to get to work and back. It would be better to live in a less efficient home closer to one's work. Similarly, while it is good to have pilot buildings of high environmental quality or having positive energy (producing more energy than they consume), it is more effective to make energy improvements of around 10% in tens of thousands of homes. The mass effect is primordial in home energy efficiency as in transport.

The energy solution depends on various parameters, so there can be no general rule. The aim however must always be to reduce the consumption of energy produced by fossil fuels. For a new house, for example, it is worth installing a passive solar panel that will provide hot water and some underfloor heating. In renovations, the use of heat pumps can often be advantageous but the choice of the cold source will depend on local conditions. For instance, an air–air heat pump can be used in the Paris region where the temperature rarely falls below $-15\,°C$, but in colder regions a geothermal pump, taking energy from underground, would be more appropriate.

Attention must be paid to insulation, exposure, lighting and household appliances if realistic energy economies are going to be made. The consumer has a strategic role to play in reducing energy consumption in an individual home.

7.5   Transport

Transport – especially road transport – is still expanding rapidly. In 1950, there were about 50 million vehicles in the world; by 2006, there were around 800 million. Speed, too, is at a premium: in 1880 it took 60 days to cross the Atlantic, today it can be done in 7 h.

It is difficult to foresee exactly what transport will be like in the future, because technological breakthroughs can change the solutions that we imagine today. Current thinking is that hydrogen will be the future energy vector associated with fuel cells of the PEMFC type. That presupposes two breakthroughs: one of the fuel and one on the means of using it. The interim solution is to improve current fuels before using them.

Even if the fuel cell and hydrogen are potential long-term solutions, they are still a long way off. Several decades of research are needed to reduce the cost of the fuel cells and improve their reliability. Hydrogen needs new infrastructures for its manufacture, storage and transport, which will take a long time to put in place. As mentioned earlier, the use of fuel cells in vehicles also creates problems of heat evacuation, operation at low temperatures and supply of catalysts, etc.

Bearing in mind the extensive experience of vehicles using internal combustion engines acquired over more than a century, it may be that the best solution for the next 10–20 years or possibly longer is the *hybrid vehicle* in which a petrol or diesel engine is used in conjunction with an electric motor running from a battery that can be recharged through the mains, with a back-up charge from the combustion engine. Hybrid vehicles already exist (more than one million Toyota Prius have been sold worldwide) but the battery is the weak point since it cannot be recharged from the mains. In this car the battery is charged by the engine and the process of braking also charges the battery. It improves the energy management, based on the use of the battery as a back-up, which enables this car to reduce its energy consumption. In town, with traffic congestion, fuel consumption is halved compared with a petrol vehicle of comparable size and power, and can easily average 5 l/100 km. With rechargeable hybrid vehicles using batteries that provide power for about 40 km, most journeys could be made using electrical energy, since in most cases commuter journeys are no longer than this.

In France, one could imagine a medium-term scenario for reducing greenhouse gases based on the use of hybrid vehicles, a small proportion of biofuels and on electricity entirely produced by hydro and nuclear energy.

Although hydro and nuclear power stations produce around 90% of French electricity, 10% is derived from burning fossil fuels to meet peak demand, which leads to an average carbon dioxide emission of 72 g/kWh or around 30 Mt of CO_2/year. On the other hand, transport is responsible for nearly a quarter of French energy consumption (around 50 MTOE) and depends entirely on oil.

The electric vehicle has not yet caught on because of its low range (around 100 km). Also its speed is lower and its price is around 15% higher than that of a normal car. On the other hand, it is quiet and non-polluting (the pollution is made by the generation of electricity that of course can be very polluting). The cost of using these cars is low: they need around 15 kWh for 400 km, or less than two euros. If all French vehicles were electric, some 75 TWh of electricity would have to be generated, representing the output of about 10 nuclear reactors of 1 GWe. A technological breakthrough in battery design could however give the electric vehicle much wider use.

With a hybrid car, the electric motor is used for short journeys in town. On the open road, either motor can be used according to the conditions and the state of the batteries.

Biofuels can never replace the entire amount of oil used by transport because there has to be a choice between driving and eating. However, some can be produced and this results in a reduction of greenhouse gas emissions. Biomass can be gasified and artificial fuels can be made by hydrogenating these gases using the Fischer–Tropsch process. This method requires, for the same quantity of biofuel produced, a smaller cultivated area than the traditional method, although part of the energy is transferred to the manufacture of hydrogen. For current levels of transport, 100 Mt of biomass would be needed plus 250 TWh of electricity to produce the hydrogen. This corresponds to 35 nuclear reactors and 20 million hectares (in France there are 18.3 million hectares of arable land). Using fallow land (1.4 million hectares) and an equivalent area of arable land, 20% of vehicle fuel could be produced and therefore greenhouse gases could be reduced by 20%. In this case, the energy from seven nuclear reactors would be needed to produce the hydrogen.

If vehicles recharged their batteries on the electric mains for 30% of their energy needs, around three additional nuclear reactors would be needed.

With seven additional nuclear reactors, all French electrical requirements would be met without having to use fossil fuel power stations to meet peak demand.

With ten additional nuclear reactors, it would be possible to meet all France's electrical needs and supply part of the energy requirements of hybrid vehicles. During off-peak hours, they would recharge their batteries and it would be possible to manufacture the hydrogen necessary for the treatment of biomass and the production of synthetic fuels.

With this solution, no more carbon dioxide would be emitted for the production of electricity, and oil consumption and road vehicle emissions would be halved.

7.6 Thermonuclear fusion

The solar energy that reaches us from the sun comes from nuclear fusion reactions taking place in its interior. The force of gravitation in the star's core is such that the densities and temperatures needed to set off thermonuclear reactions are available. Scientists have been trying for decades to induce similar fusion reactions for the production of energy, but this was only successful with the H-bomb. Fusion is not yet safe to produce large quantities of energy for industrial needs. This objective will perhaps be attained during the course of the next century, if research currently being undertaken at an international level is successful.

The fusion of two light nuclei, such as deuterium and tritium, releases more energy per nucleon than fission. Thus, the reaction (D–T) that describes the fusion of deuterium (D = $_{1}^{2}$H) and tritium (T = $_{1}^{3}$T):

$$_{1}^{2}\text{H} + _{1}^{3}\text{H} \rightarrow _{2}^{4}\text{He} + n + 17.6\,\text{MeV} \qquad (7.1)$$

releases 17.6 MeV. The nucleus of helium ($_{2}^{4}$He) takes 3.5 MeV and the neutron (*n*) 14.1 MeV. The energy released in fission is 0.85 MeV/nucleon ($= 200/236$) whereas that released in fusion is 3.5 MeV/nucleon – four times larger.

Other fusion reactions are possible. The reaction D–D can occur by two routes of equal probability:

$$_{1}^{2}\text{H} + _{1}^{2}\text{H} \rightarrow _{2}^{3}\text{He}(0.8\,\text{MeV}) + n(2.5\,\text{MeV}) \qquad (7.2)$$

$$_{1}^{2}\text{H} + _{1}^{2}\text{H} \rightarrow _{1}^{3}\text{H}(3\,\text{MeV}) + _{1}^{1}\text{H}(1\,\text{MeV}) \qquad (7.3)$$

They are more difficult to achieve than the (D–T) reaction because the probability of interaction is around 100 times smaller between 1 and 100 keV, which explains the preference for the (D–T) system.

Deuterium is a stable isotope of hydrogen present on earth. Tritium on the other hand is a radioactive nucleus not existing in the natural state with a half-life of 12.3 years. It is synthesised from ^{6}Li by the reaction:

$$^6\text{Li} + n \rightarrow \, ^4_2\text{He} + \, ^3_1\text{H} \tag{7.4}$$

Fusion is much more difficult to achieve than fission. In fission, neutrons can easily approach and penetrate uranium nuclei because there is no electric charge, but it is much more difficult to make two light nuclei fuse because both are positively charged and repel each other. This does however occur in the Sun (but very slowly[5]) where hydrogen is transformed into helium at temperatures of 10 to 15 million degrees. This fusion is called *thermonuclear* because it is induced by the kinetic energy of thermal agitation of the nuclei. There are attempts to harness this phenomenon to produce energy far into the future.

Thermonuclear fusion is achieved within a deuterium–tritium plasma.[6] For thermonuclear fusion to occur, the following conditions must be met:

- A very high temperature to enable the ions to overcome the electrostatic repulsion keeping them apart. The temperature necessary for D–T fusion is around 100 million degrees Kelvin (10^8 K).
- High particle density to ensure a large number of collisions between the ions.
- A long confinement time, so that the density and temperature of the plasma remain high and a sufficient quantity of fuel fuses.
- Confinement without contact because no material can resist such extreme conditions.

The balance of power of a fusion reactor is positive if it satisfies the Lawson criterion. This translates, for the D–T fuel heated to 10^8 K, by the equation $n \times \tau_E > 10^{20}$ s/m^3, where n is the particle density in the plasma and τ_E the energy confinement time. The objective of the research is to achieve *ignition*, in other words a self-sustaining reaction. The condition of ignition is given by $n_0 T_0 \tau_E > 6 \times 10^{22}$ (m^{-3} MJ K s), where n_0 and T_0 are respectively the density and temperature at the centre of the plasma and τ_E the energy confinement time. Two routes are possible to achieve thermonuclear fusion:

1. The *Tokamak*, originally developed in the USSR, consists in confining the hot plasma within a torus by a magnetic field that prevents it from touching the walls of the chamber. This route would appear to be the most promising for developing an industrial fusion reactor. The research is international, and the International Thermonuclear Experimental Reactor (ITER), currently being built at Cadarache in France, will replace the Joint European Torus (JET) that is the biggest tokamak to be built so far.[7]

---

[5] Fortunately, or else the Sun would quickly exhaust its fuel.

[6] A plasma is a state of matter in which electrons are detached from atoms, resulting in a cloud of electrons and ions.

[7] Its plasma volume is 140 m^3 and its performance is such that $n_0 T_0 \tau_E = 9 \times 10^{20}$.

2. *Inertial confinement fusion* for that the most promising route is to use high-energy laser beams to compress and heat D–T fuel enclosed in glass microspheres. The inertial confinement results in a higher particle density during very short periods of time. This process is more suitable for the simulation of nuclear weapons than energy production. Even though more energy is produced at the level of the microsphere than is injected, the efficiency of the laser bundles is so low (only a few per cent) that in total more energy is consumed than is produced.

In a fusion reactor based on magnetic confinement, 80% of the energy produced is taken up by the neutrons that are absorbed by a blanket surrounding the reactor. This blanket contains lithium, enabling tritium to be produced that can be used as a fuel. The blanket needs to be about a metre thick, because the fusion neutrons are very energetic (14.1 MeV). They heat the blanket as they slow down, and this energy is recovered through a coolant.

The amplification factor Q is the ratio between the energy produced by the fusion reactions and the injected external power. If $Q > 1$, the fusion reactions will provide more energy than was injected. *Break even* describes a state where $Q = 1$, when the energy produced by the fusion reaction is equal to the energy injected into the plasma. In this situation, α particles provide part of the plasma energy. In a fusion reactor, the aim is to achieve ignition, the situation when the energy from the fusion reactions compensates for the losses and there is no need to supply further energy. The plasma burns like a candle and continues burning until the fuel is exhausted. When ignition occurs, Q is infinity, because the external energy is zero.

Wastes produced during fusion are essentially the activation products of the enclosing structures that have the advantage of having short lives: they are harmless after about 100 years. On the other hand, the tritium must be carefully contained because it can easily be dispersed into the environment. The advantage of a fusion reactor is that it cannot get out of control and can be stopped very quickly.

As far as reserves are concerned [85], deuterium is fairly abundant. In seawater, there are 33 g/m^3, which translates into reserves of 4.6×10^{13} tonnes. At current rates of energy consumption, this would be sufficient to outlast the life of our planet (5 billion years). The same is not true of lithium. Natural lithium is made up of 92.5% ^7Li and 7.5% ^6Li. Its concentration in the Earth's crust is around 30 g/m^3, and reserves are estimated at 12 Mt, or three times as much as uranium. Lithium is also found in seawater (uranium also, incidentally) at a concentration of 0.17 g/m^3 that gives reserves of 240,000 tonnes. A 1000 MWe fusion reactor would consume in a year 100 kg of deuterium, 150 kg of tritium and 350 kg of lithium in the blankets to breed the tritium. There are terrestrial lithium reserves for about 5000 years. There would be enough lithium extracted from seawater to last several million years, which is still short compared to the Earth's life expectancy. D–D fusion will therefore need to be developed and brought into use.

ITER is a major research establishment. Its role is not to produce electricity but to master the conditions for creating a thermonuclear fusion plasma. The ITER reactor will be eight times bigger than the current JET international reactor, based in the UK. Whereas JET has a gain of 1, ITER will have a gain of 10, but this will still be insufficient for industrial electricity generation that requires a gain of 40. ITER's power is 500 MW_{th} and the plasma will be confined for a maximum of 400 s. The original project, 20 years ago, aimed at a power of 1500 MW_{th} and a confinement of 1000 s. The JET international machine, in England (currently the world's largest tokamak), and the Euratom Tore Supra (the world's largest superconducting tokamak) situated at Cadarache in the Rhone Valley in France served as models in the preparation of the ITER project.

Several stages are still needed before it will be possible to build an industrial fusion reactor to generate electricity for consumers. Energy must be produced reliably and continuously and at an economically competitive cost. First, a prototype reactor, currently called DEMO, with a power of 2000 MW_{th}, will enable a gain of 40 to be made and allow further studies into the production of tritium from blankets, the extraction of energy, etc. Finally, a third reactor, called PROTO, with a power of 1000 MWe, will become an industrial prototype. Bearing in mind the duration of each project (several decades) and their cost, it seems unlikely that large-scale electricity generation could be started before the next century. However, if there was international consensus that a new form of energy was essential, the development time could no doubt be shortened.

7.7   Energy storage

A technological breakthrough in this area could completely change the energy landscape. Currently, the largest amounts of energy are stored in hydro barrages through pumping water uphill with off-peak energy. This allows several GW to be stored. Large-scale compressed air storage is another solution.

Batteries can certainly be improved, but progress is likely to be marginal rather than a breakthrough. Al–air or Zn–air fuel cells may have a future using Al or Zn as energy vector. Hydrogen is also, as we have seen, an attractive means of storage, especially for renewable energies.

Superconducting magnetic energy storage (SMES) systems already exist enabling high-quality electrical power of 1–10 MW to be supplied. Their principle is to store energy in the magnetic field created by a current circulating in a superconducting magnet that is kept at the temperature of liquid helium in a cryostat, which consumes energy. Research is being carried out into the development of systems from 10 to 100 MW.

There are also two more futuristic methods of energy storage that are worth describing briefly.

The energy released in elementary nuclear reactions is measured in MeV, a million times higher than for chemical reactions. The question arises whether it would be possible to have some kind of 'nuclear' storage. A solution could be found in nuclei that possess states of high spin or isomeric configurations. Some nuclei have states situated a few MeV above the fundamental state and live long enough to be used. For instance, the nucleus of ^{178}Hf has a half-life of 36 years in a state situated 2.4 MeV above the fundamental state. Another, ^{198}Re, has a life of 300,000 years and would be capable of storing 1 TJ/l for several thousand years [86]. While this means of storing energy is understood, recovery on demand is more problematic. This would require being able to modulate the half-life of the nucleus. If this could be done, and if the process could be generalised to other nuclear excitations, which is far from evident, it would also be possible to resolve the problem of nuclear waste by reducing the half-life of long-life isotopes.

Antimatter does not exist in the world in the natural state and therefore cannot be a source of energy, but it could be used to store energy. However, in the present state of research, it would be necessary to provide infinitely more energy (around 10^8 times more) to produce the antimatter that could be later recovered.[8] The storage of antimatter in magnetic bottles also requires energy and could be dangerous if there was a confinement breakdown. And even if it were possible to store large quantities of antimatter, there would still be the problem of recovering the energy released in the form of highly energetic photons.

7.8 Negawatt-hours

Energy is abundant and cheap in developed countries. This leads to waste. However, large number of people are still in a state of energy poverty. An Ethiopian, for example, consumes on average 320 times less electricity than a French person. Sustainable development requires control of energy consumption and economies. Providing the same service with a smaller amount of energy should be the constant objective. These watt-hours that are not consumed may be described as negawatt hours.[9] These are the least polluting since they were never produced.

There are now electronic gadgets that help manage energy systems intelligently and make substantial economies. Their falling cost should enable them to be used more frequently.

Some estimates of the quantity of energy that can be saved are given below. Many other examples can be found in Reference 87.

An estimate of the energy needs for heating and air-conditioning of a standard dwelling of 100 m^2 and 250 m^3 volume is 14,300 kWh/year. If the

---

[8] The production of a billionth of a gram (10^{-9} g) in 10 years at CERN cost several hundred million euros.

[9] The term negawatt is also used but this refers to a power. It does not give an idea of the quantity of energy saved. A low energy lamp bulb of 20 W that replaces an ordinary 100-W bulb for example creates 18 negawatts but does not give a real idea of the economy achieved, since a 100-W bulb only consumes energy when it is switched on.

house is built facing the right direction, some 34% of the energy can be saved (9420 kWh) and the same house ecologically designed would use 65% less energy (5070 kWh) [87].

For heating a house, an electric convector heater converts 1 kWh of electricity into heat, whereas a heat pump can produce, according to the external temperature, around 3–4 kWh of heat for 1 kWh of electricity consumed [87].

A traditional fireplace has an efficiency of between 10% and 15%. With a simple heat exchanger, efficiency can reach 25% and up to 50% with a reverse exchanger. Closed fireplaces for new buildings or insets for existing fireplaces have an efficiency as high as 70–80%. There also exist high-performance stoves that are less polluting and have an efficiency of between 70% and 80% as opposed to 40% for older models. Wood burning boilers[10] today can be almost as efficient as heating oil or gas boilers (85%) [87].

Solar thermal energy is not yet sufficiently used despite its many advantages. It can be harnessed through special glass and well-designed verandas, wall-mounted panels, solar hot-water heaters or underfloor heating. Individual micro-generation of electricity can also be a solution to energy needs by modulating the proportion of heat and electricity generated.

Lighting in the residential-tertiary sector account for 14% of the consumption of electricity in France. Traditional incandescent light bulbs have a very low efficiency (5%) with 95% of the energy being lost in the form of heat.[11] The effectiveness of lighting can be measured in lumens per power unit. An incandescent light bulb has an output of 13 lumens/W, a halogen lamp 14 lumens/W, long-lasting compact fluorescent bulbs 60 lumens/W and fluorescent tubes 63 lumens/W. Electroluminescent diodes are another possible future solution for lighting.

The choice of domestic appliances is important in achieving energy economy. Cold appliances (refrigerator, freezer) represent the largest energy drain for a family,[12] followed by cooking appliances (hotplates and electric oven). Paradoxically, using a dishwasher with cold water intake uses less energy (20% less) and much less water (40–75% less) than washing up by hand [87].

Figure 7.1 shows the annual consumption of some domestic appliances used in a French home.

Finally, there is the problem of keeping appliances on standby (television sets,[13] video recorders, etc.). As they are on around the clock, their consumption is far from negligible. With an estimated total of around 50 W per household, it amounts to around 10 TWh/year of electricity – the power generated by 1.5 nuclear reactors.

---

[10] Oak for instance yields more energy than poplar (1700 kWh/m^3 as against 1400 kWh/m^3).

[11] Including the efficiency in electricity production (33%), transport losses and light bulb efficiency, the total efficiency of electric lighting is only around 1.5%.

[12] An American style refrigerator (with ice cube maker) uses four times as much electricity as a European model.

[13] Watching a television with a power of 80 W for 3 h/day uses 240 Wh. If it is left on standby mode (15 W) for the remaining 21 h, it will use an additional 315 Wh.

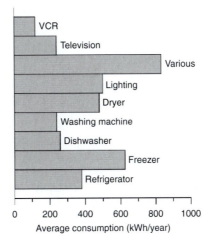

Figure 7.1   Average annual consumption of typical appliances in a French home. By introducing low-energy appliances, the total consumption could be virtually halved. [Source: www.ademe.fr]

7.9   Conclusion

Human beings have sufficient imagination and resources to develop new sources of energy when they are needed. The period that we are currently living in, where energy is abundant and cheap, unfortunately does not encourage economy or long-term research investment. It is clear that energy will become increasingly expensive in the future, and that long established habits must be changed to adapt to the new situation.

All human beings must have access to energy because it is the motor of economic development. For this to happen, energy needs to be cheap and its use should have a minimal effect on the environment. The cost of energy is a major limitation because, out of the 6 billion people in the world, 2.8 billion have less than $2 a day to live on.

The renewable energies, which humans have used since they discovered how to control fire some 500,000 years ago, are on the whole still too expensive, unreliable and too low powered. Fossil fuels (coal, oil, gas) on which our modern civilisation depends will gradually become scarcer, and other sources of energy will have to be found to replace them. Nuclear energy (uranium or thorium fission, deuterium–tritium fusion) could supply energy for tens of thousands of years. Renewable energies (provided energy needs are not too high) would be capable of supplying enough energy to last the 5 billion years before the Earth disappears. Deuterium–tritium fusion, once controlled and developed, could supply considerable quantities of high-power energy.

Better utilisation of energy, increasing conversion efficiency, and the selection and better management of energy systems is indispensable for future sustainable development.

In the future, we will need more electricity to meet new needs. For example, heat or cold could be supplied using heat pumps using the temperature source best adapted to requirements (air, ground or water), or rechargeable hybrid vehicles would enable peak demand for electricity to be smoothed by being mainly recharged at night. Moving part of our energy consumption onto electricity is only attractive when it is produced without release of greenhouse gases, that is, with nuclear, renewable energies or possibly fossil fuels if the carbon dioxide released can be captured and stored. Carbon capture on an industrial scale is far from being perfected, and it would take thousands of plants like the existing prototypes (1 Mt of CO_2/year) to sequestrate a significant proportion of the carbon dioxide currently released by human activities.

In the more distant future, it will be important to make best use of the carbon atoms extracted from nature (remaining fossil fuels or biomass). In particular, they need to be used to manufacture liquid fuels for transport and other purposes where they are indispensable. And they should not be burnt to provide energy for transformation processes and should be prevented from releasing carbon dioxide. This means that external heat sources that do not release greenhouse gases (renewable or nuclear) should be used, and that hydrogen, produced from water by electrolysis using clean electricity, should be harnessed to achieve balanced chemical reactions.

Bibliography

[1] SMIL V., *Energies*, MIT Press, 1998.

[2] ROQUES-CARMES C. and LEFÉBURE N., *L'énergie et ses secrets*, Fernand Nathan Éditeur, 1984.

[3] *Informations utiles*, CEA, 2006 and previous editions.

[4] GRIMM G., *Les nouvelles de l'environnement*, no. 57.

[5] *Mémento sur l'énergie*, CEA, 2006.

[6] ARQUÈS Ph., *La pollution de l'air, causes, conséquences, solutions*, Édisud, 1998.

[7] PHILLIPS A.C., *The Physics of Stars*, John Wiley & Sons, 1994.

[8] PHILLIPS K.J.H., *Guide to the Sun*, Cambridge University Press, 1992.

[9] REEVES H., *Patience dans l'azur, l'évolution cosmique*, Éditions du Seuil, 1981.

[10] LE TREUT H. and JANCOVICI J.M., *L'effet de serre*, Dominos, Flammarion, 2001.

[11] DGEMP, *Chiffres clés de l'énergie*, 2006 edition, www.developpement-durable.gouv.fr/IMG/pdf/Repere.pdf.

[12] *L'environnement en France*, Ifen, Éditions La Découverte, 1998.

[13] LEYGONIE R., MARTIN B. and PRILLIEUX M., *Cette roche nommée pétrole*, Éditions Belin, 1983.

[14] MÉRENNE-SCHOUMAKER B., *Géographie de l'énergie*, Nathan Université, 1993.

[15] ROQUES-CARMES C. and LEFÉBURE N., *L'énergie et ses secrets*, Fernand Nathan Éditeur, 1984.

[16] *Environnement, énergie, santé*, Brochure de la Société Française d'Énergie Nucléaire, groupe Languedoc-Roussillon, June 1997.

[17] GABBARD A., *Coal Combustion: Nuclear resource or danger*, Oak Ridge National Laboratory, 2001.

[18] PRÉAT A., *L'énergie d'aujourd'hui et de demain: le point de vue d'un géologue?*, Université Libre de Bruxelles.

[19] Union Française des Industries Pétrolières, www.ufip.fr.

[20] International Tanker Owner Pollution Federation, www.itopf.com.

[21] *Enerpresse*, no. 7425, 1999.

[22] *BP Statistical Review of World Energy*, 2007 and previous editions.

[23] *CH4*, Revue du Gaz de France, 1999.

[24] Gaz de France, www.gdf.fr.

[25] *Enerpresse*, no. 7422, 1999.

[26] CHARPIN J.M., DESSUS B. and PELLAT R., *Étude économique prospective de la filière électrique nucléaire*, La documentation française, 2000.

[27] CAMPBELL C. et LAHERRÈRE J., *Pour la science*, no. 247, May 1998.

[28] Commissariat général du plan, Club 'Énergie, prospective et débats', Rapport d'activité 2000.

[29] BIRRAUX C. and LE DÉAUT J.Y., *L'état actuel et les perspectives techniques des énergies renouvelables*, Rapport de l'Office parlementaire d'évaluation des choix scientifiques et technologiques, 2002.

[30] VERNIER J., *Les énergies renouvelables*, *Que sais-je?*, Presses Universitaires de France, 1997.

[31] PHILLIPS K.J.H., *Guide to the Sun*, Cambridge University Press, 1992.

[32] CUNTY G., *Éoliennes et aérogénérateurs*, Edisud, 2001.

[33] NOËL J.M., *Informations sur l'énergie éolienne*, UEEGM/Saclay, DESS EPE/GNTE, 2/56-1, 2000–2001.

[34] RAVOIRE J., *Sur l'énergie et l'environnement*, unpublished document, September 2001.

[35] *Comité Biomassse-énergie*, Commissariat à l'énergie solaire, May 1980.

[36] BOBIN J.L., NIFENECKER H. and STÉPHAN C., *L'énergie dans le monde, bilan et perspectives*, EDP Sciences, 2001.

[37] GOGUEL J., *La géothermie*, Doin Éditeurs, 1975.

[38] BAUQUIS P.R., 'Un point de vue sur les besoins et les approvisionnements en énergie à l'horizon 2050', *Revue de l'énergie*, no. 509, 1999.

[39] Société Française de Radioprotection, www.sfrp.asso.fr.

[40] REUSS P., *La neutronique*, *Que sais-je?*, 1999.

[41] TELLIER H., *Cinétique des réacteurs nucléaires*, CEA, INSTN, 1993.

[42] CAVEDON J.M., *Réacteurs nucléaires*, Électronucléaire, une présentation par des physiciens, Publication du cercle d'études sur l'énergie nucléaire, 1999.

[43] BARRÉ B., *Actes des journées de l'énergie au Palais de la Découverte*, 2001.

[44] ELECNUC, *Les centrales nucléaires dans le monde*, CEA, 2001 edition.

[45] TURLAY R. (ed.), *Les déchets nucléaires, un dossier scientifique*, EDP Sciences, 1997.

[46] WIESENFELD B., *L'atome écologique*, EDP Sciences, 1998.

[47] CAVEDON J.M., *La séparation isotopique*, Électronucléaire, une présentation par des physiciens, Publication du cercle d'études sur l'énergie nucléaire, 1999.

[48] *L'industrie nucléaire et ses marchés en Europe, Perspectives stratégiques et financières*, EUROSTAF, 1996.

[49] BATAILLE C. and GALLEY R., *L'aval du cycle nucléaire*, Office parlementaire des choix scientifiques et technologiques, Rapport de l'Assemblée Nationale no. 1359, du Sénat no. 195, 1999.

[50] BATAILLE C., *Les possibilités d'entreposage à long terme de combustibles nucléaires irradiés*, Office parlementaire des choix scientifiques et technologiques, Rapport de l'Assemblée Nationale no. 3101, du Sénat no. 347, 2001.

[51] LANGRAND J.C., Service d'études économiques, CEA.

[52] *Uranium 1999, Ressources, production et demande*, OCDE, Agence pour l'Énergie Nucléaire, 2000.

[53] BASTARD P., FARGUE D., LAURIER P., MATHIEU B., NICOLAS M. and ROOS P., *Électricité, voyage au cœur d'un système*, Eyrolles, 2000.

[54] CONAN J., *Actes des journées de l'énergie au Palais de la Découverte*, May 2001.

[55] *A Guide to Cogeneration*, European Association for the Promotion of Cogeneration, 2001, www.cogeneurope.eu/wp-content/uploads//2009/02/educogen_cogen_guide.pdf.

[56] DOUAUD A., *Actes des journées de l'énergie au Palais de la Découverte*, May 2001.

[57] DAMERY E., note technique CEA/DTEN no. 2002–11.

[58] WISER W.H., *Energy Resources*, Springer, 1999.

[59] ASKILL J., Millikin University, 1999.

[60] *L'hydrogène*, Les dossiers de l'énergie no. 14, Ministère de l'industrie, du commerce et de l'artisanat.

[61] COMTE A., 'Les piles à combustible, principes et quelques applications', *Auvergne Sciences*, no. 50, April 2001.

[62] Réseau Paco, www.reseaupaco.org.

[63] BIENVENU C., *Vous avez dit énergie?* Sofedir, 1980.

[64] JOUSSAUME S., *Climat d'hier à demain*, CNRS Éditions, 1999.

[65] NESME-RIBES E. and THUILLIER G., *Histoire solaire et climatique*, Belin, 2000.

[66] GIEC (Group d'Experts Intergouvernemental sur l'Evolution du Climat), 2001.

[67] *Le diesel et son environnement*, CLID, March 1999.

[68] CITEPA (Centre Interprofessionnel Technique d'Etudes du la Pollution Atmosphérique), 1998.

[69] Conférence à l'école des Mines de Paris, 1997.

[70] Laboratoire d'hygiène de la Ville de Paris, 1997.

[71] MONNA F., *La Recherche*, no. 340, March 2001, p. 50.

[72] Ministère de l'aménagement du territoire et de l'environnement, www.environnement.gouv.fr.

[73] *La radioactivité naturelle en 10 épisodes*, Société Française de Radio-protection, 1998.

[74] OCDE-AEN and Conseil Scientifique des Nations Unies and www.in2p3.fr.

[75] *Gazette Nucléaire*, info IPSN no. 158, 1994.

[76] PÉLLISSIER-TANON J., *Le jaune et le rouge*, 1999.

[77] GR21 (Groupe de réflexion énergie environnement au XXIème siècle), SFEN, 2001.

[78] GALLE P., *Toxiques nucléaires*, Masson, 1997.

[79] SYROTA A., 'Les nuisances de l'énergie sur le vivant : l'influen-ce des faibles doses', *Actes des journées de l'énergie au Palais de la Découverte*, 2001.

[80] Encyclopedia Universalis.

[81] Jauron R. and Laroche P., Chrysotile Laboratory.

[82] Mandil C., 'Les hydrocarbures, une énergie d'avenir', *Actes des journées de l'énergie au Palais de la Découverte*, 2001.

[83] Santos I., Cité des Sciences de la Vilette, www.cite-science.fr.

[84] Lucciani J.F., CEA, 2001.

[85] Informations CEA.

[86] Gogny D., *La matière dans tous ses états*, École Joliot-Curie, 1986.

[87] Salomon T. and Bedel S., *La maison des négawatts*, Terre Vivante, 2001.

[88] www.wikipedia.fr

[89] Lambert G., Chappellaz J., Foucher J.P. and Ramstein G., *Le méthane et le destin de la Terre*, EDP Sciences, 2006.

[90] Bonal J. and Rossetti P., *Énergies alternatives*, Omniscience, 2007.

[91] International Energy Agency, www.iea.org

[92] Durand B., *Énergie et environnement*, EDP Sciences, 2007.

[93] Bouillon H., Fösel P., Neubabth J. and Winter W., *Wind Report 2004*, E.ON Netz, 2004.

[94] Réseau de Transport d'Électricité, www.rte-france.com

[95] Association Français de l'Hydrogène, www.afh2.org

[96] Comité des Constructeurs Français d'Automobiles, www.ccfa.fr

[97] Rojey A., IFP, Conférence à ECRIN, September 2007.

[98] Commissariat à l'Énergie Automique, www.cea.fr/english-portal

Index

Page numbers followed by f and t indicate figures and tables, respectively.